Springer Optimization and Its Applications

VOLUME 83

Aims and Scope
Optimization has been expanding in all directions at an astonishing rate during the last few decades. New algorithmic and theoretical techniques have been developed, the diffusion into other disciplines has proceeded at a rapid pace, and our knowledge of all aspects of the field has grown evenmore profound. At the same time, one of the most striking trends in optimization is the constantly increasing emphasis on the interdisciplinary nature of the field. Optimization has been a basic tool in all areas of applied mathematics, engineering, medicine, economics, and other sciences.

The series Springer Optimization and Its Applications publishes undergraduate and graduate textbooks, monographs and state-of-the-art expository work that focus on algorithms for solving optimization problems and also study applications involving such problems. Some of the topics covered include nonlinear optimization (convex and nonconvex), network flow problems, stochastic optimization, optimal control, discrete optimization, multiobjective programming, description of software packages, approximation techniques and heuristic approaches.

For further volumes:
http://www.springer.com/series/7393

Pavel S. Knopov • Olena N. Deriyeva

Estimation and Control Problems for Stochastic Partial Differential Equations

Springer

Pavel S. Knopov
Department of Mathematical Methods
of Operation Research
V.M. Glushkov Institute of Cybernetics
National Academy of Sciences
of Ukraine
Kiev, Ukraine

Olena N. Deriyeva
Department of Mathematical Methods
of Operation Research
V.M. Glushkov Institute of Cybernetics
National Academy of Sciences
of Ukraine
Kiev, Ukraine

ISSN 1931-6828
ISBN 978-1-4939-4493-4 ISBN 978-1-4614-8286-4 (eBook)
DOI 10.1007/978-1-4614-8286-4
Springer New York Heidelberg Dordrecht London

Mathematics Subject Classifications (2010): 35R60, 60H15, 93C20

Printed on acid-free paper

Springer is part of Springer Science+Business Media (www.springer.com)

Preface

In classical physics and engineering many problems are described by ordinary or partial differential equations. One usually considers an idealized situation, namely, the uncontrolled perturbations of a random system are excluded, although from the practical point of view taking into account not only the random evolution but also external random perturbations is of great interest.

To illustrate the motivation given above, consider the following situation. Let the evolution system be described by the differential equation

$$Lu + f(u) = 0,$$

Where L is some differential operator.

The solution to this equation (provided that it exists) is a deterministic function which completely characterizes the system. Suppose now that the system is subject to small external random perturbations. Under certain conditions it can be proved, using the central limit theorem, that the system is perturbed by a Gaussian white noise. Therefore, in a somewhat simplified situation, we can assume that the evolution system is described by the differential equation

$$Lu + f(u) = \xi,$$

where ξ is a random Gaussian perturbation.

The solution to this equation is no longer a deterministic function, and therefore the behavior of the system cannot be absolutely precise. However, it is possible to predict (in some sense optimally) the behavior of the system using more simple observations of the equation. For systems described by ordinary stochastic differential equations, this method allows to approach many practical problems in a new way, based on the new theory of automatic control of stochastic systems. But there are many problems, both theoretical and applied, solving of which requires new approaches. This often happens due to the fact that the investigated processes are described by stochastic partial differential equations. This problem is extremely extensive, so one cannot describe the whole range of applications associated with it.

To illustrate this let us focus on two problems that are important both from theoretical and applied points of view.

1. *The hydrodynamic instability problem.* In the modern approach for solving this problem, a fluid is considered as a nonlinear mechanical system with a very large number of degrees of freedom. Due to the nonlinearity of the dynamics various degrees of freedom mutually interact. In Eulerian description of the fluid motion such interactions are the inertial interactions between the inhomogeneities of the velocity field; the interaction constant is called the Reynolds number. Hence, for sufficiently large Reynolds numbers (for which the fluid motion may become unstable), the interactions between the degrees of freedom are very strong. Therefore, an unstable perturbation of a single degree of freedom rapidly leads to the perturbation of many other degrees of freedom, and the fluid motion becomes very complex and irregular, hardly amenable to the particular description. It is reasonable to describe this motion only statistically, using the methodology of the theory of random fields. And there, of course, arises the problem how to predict such a field based on the observations of another field with more simple structure.

   Hence, it becomes possible to obtain statistical estimates for the values of the process arising in many natural and technological phenomena such as the formation of sea waves, the resistance crisis when a fluid flows around curved profiles, and thermal convection.
2. *The problem of the optimal control of technological processes.* Consider the problem arising in the study of gases' absorption and desorption. Let a tube of length L be filled with an absorbing material (sorbent). We choose the tube axis as the coordinate axis, and assign the tube entrance, through which the gas–air mixture is supplied starting from the time moment $t = 0$, as the origin. If $u(x, t)$ is the gas concentration at time t in the tube layer x, then we have the following relation for the case of the low concentration of the supplied gas:

$$u_{kt} + \frac{\beta}{\nu} u_t + \beta\gamma u_x = 0,$$

   where β is the kinetic coefficient, ν is the gas velocity, and $1/\gamma$ is the Henry coefficient.

Many factors such as all sorts of irregularities in the distribution of the sorbent in the tube, the irregularity of the supply of gas mixture flow in time, its heterogeneity, and others make it necessary to consider equations with random coefficients and additional terms which reflect the diversity of random deviations from the process, which can be characterized using a multiparameter white noise. Here, again one comes across the problem of finding estimates for the parameters of the process. Among such problems there are of course the problems of existence and uniqueness of the solution to stochastic partial differential equations, which are of great theoretical and practical interest.

The book is focused on the study of the stochastic differential equations of hyperbolic type. Historically, the attention was focused first on the study of the stochastic Darboux equations, where the two-parameter Wiener field is taken as a noise. Therefore, the first question is how to construct a stochastic integral with respect to this field, and what are its basic properties. Similar to the one-parameter case, the stochastic integral was built with respect to a two-parameter martingale, which makes it possible to study the stochastic differential equations of hyperbolic type, where the two-parameter martingale was considered as noise. Such integrals and martingales are considered in Cairoli and Walsh [7], Etemadi and Kallianpur [19], Wong and Zakai [72], Gyon and Prum [36], and many others. One of the most fundamental results in the theory of stochastic differential equations is the Girsanov theorem, which provides the possibility to solve many recognition, estimation, filtration, forecasting, and optimal control problems. Such problems are considered completely enough in Gikhman and Skorokhod [26–29], Liptzer and Shiryaev [54], Novikov [63], and many other monographs. Similar problems arise for stochastic partial differential equations as well, and the two-parameter analog of Girsanov theorem [13, 50] gives us a key to find the solution for the above-mentioned estimation and control problems. On the other hand, related problems arise in the theory of evolution equations with Wiener Hilbert-valued process.

The problems described above are the basic content of this book. The authors tried to make the exposition of the material self-contained, in particular, to provide all necessary definitions and results used later on in the proofs.

Kiev, Ukraine Pavel S. Knopov
Kiev, Ukraine Olena N. Deriyeva

Acknowledgments

We are very grateful to Senior Publishing Editor Elizabeth Loew for her helpful support and collaboration in preparation of the manuscript.

We thank our colleagues from V. M. Glushkov Institute of Cybernetics of National Academy of Sciences of Ukraine for many helpful discussions on the problems and results described and presented in this book.

We thank our colleagues V. Knopova and L. Vovk for invaluable help during the preparation of our book for publication.

Contents

Chapter 1
Two-Parameter Martingales and Their Properties

This chapter provides well-known results concerning the properties of two-parametric martingales and stochastic integration on the plane. We begin with the auxiliary chapter, which also contains some facts which are of independent interest. Our standard references for the results below are [20–24, 40, 42, 44, 47, 48, 65, 71].

1.1 Definitions and Global Properties

Denote by R_+^2 the set of pairs of real numbers $z = (t, s)$, $t \geq 0$, $s \geq 0$. We write $z \leq z'$, where $z' = (t', s')$, if $t \leq t'$ and $s \leq s'$; and $z < z'$, if $t < t'$ and $s < s'$, $[z, z'] = [t, t'] \times [s, s']$.

Let $(\Omega, \mathfrak{I}, P)$ be some probability space. All random variables and random functions considered below are given on this space. We call a σ-algebra family $(\mathfrak{I}_z)_{z \in R_+}$ the flow, if $\mathfrak{I}_z \subset \mathfrak{I}$ and $z \leq z'$ implies $\mathfrak{I}_z \subset \mathfrak{I}_{z'}$. We assume also that our flow is complete, i.e., that $\mathfrak{I}_0$ contains all sets of $\mathfrak{I}$ with zero probability measure, and $\mathfrak{I}_z$ is right-continuous, which means that $\mathfrak{I}_z = \bigcap_{z<z'} \mathfrak{I}_{z'}$ holds for any z. For a given flow we define the σ-algebras: $\mathfrak{I}_z^1 = \vee_{s \geq 0} \mathfrak{I}_{(t,s)}$, $\mathfrak{I}_z^2 = \vee_{t \geq 0} \mathfrak{I}_{(t,s)}$, $\mathfrak{I}_z^* = \mathfrak{I}_z^1 \vee \mathfrak{I}_z^2$. Here $\vee_A \mathfrak{I}_z$ is the minimal σ-algebra containing all $\mathfrak{I}_z$, where z takes values in some set A.

Definition 1.1 A function $\xi = \xi(z) = \xi(z,\omega)$, $z \in R^2$, $\omega \in \Omega$, is called adapted to the flow $(\mathfrak{I}_z)$ (or $\mathfrak{I}_z$-adapted), if the random variable $\xi(z)$ is $\mathfrak{I}_z$-measurable for any $z \in R^2$.

Further we consider random fields defined on R_+^2.

Definition 1.2 A random function $\xi(z) = \xi(t,s)$, $z \in R_+^2$ is called a martingale (or two-parametric martingale, or $\mathfrak{I}_z$-martingale), if

A1. $\xi(z)$ is $\mathfrak{I}_z$-adapted;
A2. $E|\xi(z)| < \infty$ for any $z \in R_+^2$;
A3. $E(\xi(z')/\mathfrak{I}_z) = \xi(z)$ for any $z \leq z'$.

P.S. Knopov and O.N. Deriyeva, *Estimation and Control Problems for Stochastic Partial Differential Equations*, Springer Optimization and Its Applications 83,
DOI 10.1007/978-1-4614-8286-4_1,

Definition 1.3 A random function $\xi(z) = \xi(t,s)$, $z \in R_+^2$ is called a strong martingale if it satisfies conditions A1, A2, and in addition

A4. $E\left(\xi\left(z, z'\right] / \Im_z^*\right) = 0$ for any $z \leq z'$;
A5. $\left(\xi(t, 0), \Im_z^1\right)$, $t \geq 0$, and $\left(\xi(0, s), \Im_z^2\right)$, $s \geq 0$, are martingales.

In what follows we call the processes $\left(\xi(t, 0), \Im_z^1\right)$, $t \geq 0$, and $\left(\xi(0, s), \Im_z^2\right)$, $s \geq 0$, the boundary values of the martingale $\xi(z) = \xi(t,s)$.

Remark 1.1 It follows directly from the definition that a strong martingale is a martingale.

Denote by $\xi_{t\cdot}(s)$ and $\xi_{\cdot s}(t)$, respectively, the t- and s-marginals of the function $\xi(z) = \xi(t,s)$. Observe, that the processes $\left(\xi_{\cdot s}(t), \Im_z^1\right)$ and $\left(\xi_{t\cdot}(s), \Im_z^2\right)$ are one-parameter martingales. Indeed, for $s < s'$, $t > 0$, we have

$$\begin{aligned} E\left[\xi_{t\cdot}(s') / \Im_z^2\right] - \xi_{t\cdot}(s) &= E\left[\left(\xi_{t\cdot}(s') - \xi_{t\cdot}(s)\right) / \Im_z^2\right] \\ &= E\left[\xi\left((0, s), (t, s,')\right) / \Im_z^2\right] + E\left[\left(\xi_{0\cdot}(s') - \xi_{0\cdot}(s)\right) / \Im_z^2\right] = 0 \end{aligned}$$

Similar equality holds for $\xi_{\cdot s}(t)$.

Definition 1.4 A random function $\xi(z) = \xi(t,s)$, $z \in R_+^2$, adapted to the flow $(\Im_z)$, is called

(a) a weak (two-parametric) sub-martingale, if its boundary values are sub-martingales, and for any $z \leq z'$ the condition $E\left(\xi\left(z, z'\right] / \Im_z\right) \geq 0$ is satisfied;
(b) strong sub-martingale, if its boundary values are sub-martingales, and for any $z \leq z'$ the condition $E\left(\xi\left(z, z'\right] / \Im_z\right) \geq 0$ is satisfied;
(c) 1- (2-) sub-martingale, if processes $\left(\xi_{\cdot s}(t), \Im_z^1, t \geq 0\right)$ (respectively, $(\xi_{t\cdot}(s)$, $\Im_z^2, s \geq 0)$) are sub-martingales for any s (respectively, t);
(d) bi-sub-martingale, if it is 1- and 2-sub-martingales simultaneously.

Definition 1.5 A random function $\xi(z)$ is called a strong, weak, 1-, 2-, or bi-super-martingale, if $-\xi(z)$ is, respectively, strong, weak, 1-, 2-, or bi-sub-martingale.

Remark 1.2 Random functions which are simultaneously weak sub- and super-martingales are called weak martingales. In this case definitions of 1-, 2-, and bi-martingales are similar.

Remark 1.3 Apparently, a strong martingale is a bi-martingale; on the other hand, bi-martingale and martingale are weak martingales.

Remark 1.4 [22] Bi-martingale is a martingale.

Indeed, for $z < z'$ we obtain

$$\begin{aligned} E[\xi(0, z'] - \xi(0, z] / \Im_z] &= E\left[\xi\left((t, 0), (t', s) / \Im_z^1 / \Im_z\right]\right. \\ &\quad + E\left[\xi\left((t, 0), (t, s') / \Im_z^2 / \Im_z\right]\right. = 0. \end{aligned}$$

On the other hand,

$$\xi(0,z'] - \xi(0,z] = \xi(z') - \xi(z) - (\xi(t',0) - \xi(t,0)) - (\xi(0,s') - \xi(0,s)),$$

implying

$$0 = E[\xi(0,z'] - \xi(0,z]/\mathfrak{I}_z] = E[\xi(z')/\mathfrak{I}_z] - \xi(z).$$

In modern theory of two-parametric martingales one uses the special property of the flow $(\mathfrak{I}_z)$ of σ-algebras, proposed by Cairoli and Walsh, and also known as condition **F4**.

F4. For any z the σ-algebras $\mathfrak{I}_z^1$ and $\mathfrak{I}_z^2$ under given σ-algebra $\mathfrak{I}_z$ are conditionally independent: for any bounded random variables a and b which are, respectively, $\mathfrak{I}_z^1$- and $\mathfrak{I}_z^2$ -measurable, we have $E(ab/\mathfrak{I}_z) = E(a/\mathfrak{I}_z)E(b/\mathfrak{I}_z)$.

Remark 1.5 Condition F4 is equivalent to the following one: For any integrable random variable ζ the conditions below hold true:

(a) $E(\zeta/\mathfrak{I}_z) = E(\zeta/\mathfrak{I}_z^1/\mathfrak{I}_z^2) = E(\zeta/\mathfrak{I}_z^2/\mathfrak{I}_z^1)$, (1.1)

(b) if in addition $\mathfrak{I}_z = \mathfrak{I}_z^1 \cap \mathfrak{I}_z^2$, then (a) is equivalent to $E(\zeta/\mathfrak{I}_z^1/\mathfrak{I}_z^2) = E(\zeta/\mathfrak{I}_z^2/\mathfrak{I}_z^1)$.

Theorem 1.1 [23] *If the flow $(\mathfrak{I}_z)$ of σ-algebras satisfies the Cairoli–Walsh (F4) condition, then the notions of a two-parametric martingale and that of a bi-martingale coincide.*

Proof It is enough to prove that if condition (1.1) is satisfied, then a martingale is a bi-martingale. Indeed, by condition (1.1) we have

$$\begin{aligned} 0 &= E(\xi(z') - \xi(t',s)/\mathfrak{I}_{(t',s)}) = E(E(\xi(z') - \xi(t',s)/\mathfrak{I}_z^1)/\mathfrak{I}_z^2) \\ &= E(\xi(z') - \xi(t',s)/\mathfrak{I}_z^2) \end{aligned}$$

Hence,

$$E((\xi(z') - \xi(t',s))/\mathfrak{I}_z^2) = 0,$$

implying

$$E(\xi(z,z']/\mathfrak{I}_z^2) = E(\xi(z') - \xi(t',s)/\mathfrak{I}_z^2) - E(\xi(t,s') - \xi(z)/\mathfrak{I}_z^2) = 0.$$

Similarly,

$$E(\xi(z,z']/\mathfrak{I}_z^1) = 0.$$

□

Remark 1.6 Let $\xi = \xi(z)$ be a martingale defined on $[0,z^*]$. The function $\xi(z)$ is defined by the flow $(\mathfrak{I}_z)$ uniquely, almost surely for each z, and $\xi(z) = E(\xi(z^*)/\mathfrak{I}_z)$. On the other hand, for any integrable random variable γ and any flow $(\mathfrak{I}_z)$, the function $\xi(z) = E(\gamma/\mathfrak{I}_z)$ is a martingale. Generally speaking, a bi-martingale cannot be constructed in the same way starting from arbitrary random variable γ. The functions $\xi_1(z) = E(\gamma/\mathfrak{I}_z^1/\mathfrak{I}_z^2)$ and $\xi_2(z) = E(\gamma/\mathfrak{I}_z^2/\mathfrak{I}_z^1)$ are, generally speaking, not equal, not bi-martingales and not $\mathfrak{I}_z$-adapted. If the Cairoli–Walsh condition (F4) is satisfied and $\mathfrak{I}_z^1 \wedge \mathfrak{I}_z^2 = \mathfrak{I}_z$, then $\xi_1(z) = \xi_2(z)$, and this function is a martingale; hence, it is a bi-martingale.

In what follows we consider only separable random functions defined on the rectangle $[0,T]^2 := [0,T] \times [0,T]$.

Remark 1.7 Let $\xi(z) = \xi(t,s)$ be $(\mathfrak{I}_z)$-adapted, $z \in [0,T]^2$. Then for each z the function $\xi(z)$ is determined uniquely with probability 1 by the value $\xi(T,T)$ and the flow $(\mathfrak{I}_z)$:

$$\xi(z) = E(\xi(T,T)/\mathfrak{I}_z).$$

Remark 1.8 For any random variable ζ and the flow $(\mathfrak{I}_z)$, the function $\xi(z) = E(\zeta/\mathfrak{I}_z)$ is a martingale. In the general case, it is not possible to construct a bi-martingale from an arbitrary random variable ζ, since in general the functions $\hat{\xi}(z) = E(\zeta/\mathfrak{I}_z^1/\mathfrak{I}_z^2)$ and $\tilde{\xi}(z) = E(\zeta/\mathfrak{I}_z^2/\mathfrak{I}_z^1)$ are not equal, not martingales and, even more, may fail to be $(\mathfrak{I}_z)$ -adapted. If the Cairoli–Walsh condition (F4) is satisfied and $\mathfrak{I}_z^1 \wedge \mathfrak{I}_z^2 = \mathfrak{I}_z$, then $\hat{\xi}(z) = \tilde{\xi}(z)$, and this function is a bi-martingale; hence, it is a martingale.

Now we introduce some conventions which will be used throughout the book. We assume all functions under consideration to be separable. We write $\xi(z) \in L^p$, $L^p = L^p(\mathfrak{I}_z), p > 0$, if the function $\xi(z)$, $z \in [0,T]^2$ is adapted to the flow $(\mathfrak{I}_z)$ and $E|\xi(z)|^p < \infty$ for any $z \in [0,T]^2$. We denote the classes of martingales, weak martingales, bi-martingales, strong martingales, 1- or 2-martingales with respect to a given flow $(\mathfrak{I}_z)$, respectively, by M, wM, bM, sM, $1 - M$, $2 - M$. The sub-classes of those classes, consisting of p -integrable functions (that is, of functions belonging to L^p), we denote by M_p, wM_p, etc. We will add the upper index c to denote the sub-classes, consisting of continuous random functions. For example, $sM_2{}^c$ denotes the class of continuous square integrable strong martingales.

The simplest example of a martingale with continuous argument is a two-parameter Wiener process or a Wiener field $W(z)$, $z \in R_+^2$, with the following properties:

(a) $W(t,0) = W(0,s) = 0$;
(b) for any points $z_1, z_2, \ldots, z_n$, $0 \le z_1 < z_2 < \cdots < z_n$, the variables $W(z_1, z_2]$, $W(z_2, z_3]$, ..., $W(z_{n-1}, z_n]$ are mutually independent;
(c) the random variable $W(z, z + u]$, $u = (x, y) > 0$, has a Gaussian distribution with parameters $(0, xy)$.

Denote by $\Im_z = \Im_z^W$ the minimal σ-algebra of sets from Ω, with respect to which all variables $W(z')$, $z' \in (0, z]$, are measurable. Let us prove that the flow $(\Im_z)$, $z \in R_+^2$, is right-continuous. It is sufficient to show that

$$E[\eta/\Im_z] = E[\eta/\Im_{z+}]$$

for any $\eta \in H$, where H is a dense subset in the class L_1 of integrable random variables. Take a family of random variables η of the form

$$\eta = f_1 \circ W(A_1) \cdot f_2 \circ W(A_2) \cdot \ldots \cdot f_n \circ W(A_n) \cdot g_1 \circ W(B_1) \cdot g_2 \circ W(B_2) \cdot \ldots \cdot g_m \circ W(B_m),$$

where f_k and g_i are bounded continuous functions, $A_k = (z_k, z'_k]$, $k = 1, \ldots, n$ is a rectangle from $[0,z]$, and $B_j = (z''_j, z'''_j]$, $j = 1, \ldots, m$, are rectangles lying outside $[0,z]$. Denote by η_ε the random variable obtained from η by replacing the sets B_j with $B_j^\varepsilon = B_j \backslash [0, z + \varepsilon]$, $\varepsilon > 0$. The last set is always represented as a sum of two (or less) rectangles. Clearly, $\eta_\varepsilon \to \eta$ in L_1 as $\varepsilon \to 0$, $\varepsilon > 0$, $\varepsilon \in R_+^2$. For $u \le v \le u + \varepsilon$ we have

$$\begin{aligned} E[\eta_\varepsilon/\Im_{z+\varepsilon}] &= f_1 \circ W(A_1) \cdot f_2 \circ W(A_2) \cdot \ldots \cdot f_n \circ W(A_n) \\ &\quad \cdot E\left[g_1 \circ W\left(B_1^\varepsilon\right) \cdot g_2 \circ W\left(B_2^\varepsilon\right) \cdot \ldots \cdot g_m \circ W\left(B_m^\varepsilon\right)\right], \end{aligned}$$

hence, $\Im_{z+} = \Im_z$ for any $z \ge 0$. Thus, for any variable $\eta \in L^1$ we obtain $E\left[\eta/\Im_z^1/\Im_z^2\right] = E\left[\eta/\Im_z^2/\Im_z^1\right] = E[\eta/\Im_z]$, that is, in this case the flow $(\Im_z)$ satisfies the commutation conditions, and $\Im_z^1 \wedge \Im_z^2 = \Im_z$.

It is easy to see that any two-parametric Wiener process is a strong martingale.

Theorem 1.2 [23] *Two-dimensional Wiener process has a continuous modification, i.e., there exists another two-dimensional process* $\left(\widetilde{W}(z), \Im_z, z \in [0,T]^2\right)$ *such that*

(1) $P\left\{W(z) = \widetilde{W}(z)\right\} = 1$ for any $z \in [0,T]^2$;

(2) $P\left\{\omega : \widetilde{W}(z,\omega) \text{ is a continuous function}\right\} = 1.$

Taking into account the above properties, the proof of the above theorem is similar to the proof in the one-dimensional case.

Theorem 1.3 [19] *Let* $\left(X(z), \Im_z, z \in [0,T]^2\right)$ *be a continuous random field, and suppose that the Cairoli–Walsh condition F4 is fulfilled. Then the following statements are equivalent:*

(a) $(X(t), \Im_t, t \in [0,T])$ *is a Wiener field;*

(b) *for any* z ∈ D_T, $\left(X(u,s), \Im_{(u,s)}^1, u \in [0,T]\right)$ *and* $\left(X(t,v), \Im_{(t,v)}^2, v \in [0,T]\right)$ *are one-parametric Wiener processes with variable parameters* s *and* t, *respectively.*

Proof Clearly, (a) implies (b). Let us prove the reverse. It is sufficient to show that for any two non-intersecting rectangles $(z,z']$ and $(z'',z''']$ from $[0,T]^2$ the random variables $X(z, z']$ and $X(z'',z''']$ are independent, $X(z, z']$ is normally distributed with mean 0 and variance $(t' - t)(s' - s)$.

Since $\left\{X(u,s), \Im^1_{(u,s)}, u\in[0,T]\right\}$ and $\left\{X(u,s'), \Im^1_{(u,s')}, u\in[0,T]\right\}$ are one-parametric Wiener processes, $X(t',s') - X(t,s')$ and $X(t',s) - X(t,s)$ are independent from $\Im^1_{(t,s')}$ and $\Im^1_{(t,s)}$, respectively, and since, $\Im^1_{(t,s')} = \Im_{(t,s)}, X(z,z']$ does not depend on $\Im^1_z$.

We can prove similarly that $X(z, z']$ does not depend on $\Im^2_z$. Thus, $X(z, z']$ is independent of $\Im^1_z \vee \Im^2_z$, and the analogous statement holds true for $X(z'',z''']$. Moreover, since the rectangles $(z,z']$ and $(z'',z''']$ do not intersect, the variable $X(z, z']$ is measurable with respect to $\Im^1_z \vee \Im^2_z$, and $X(z'',z''']$ is measurable with respect to $\Im^1_{z''} \vee \Im^2_{z''}$. Hence, $X(z, z']$ and $X(z'',z''']$ are independent. Hence,

$$\begin{aligned}
\exp\left\{-\frac{u^2}{2}(t'-t)s'\right\} &= E\exp\{iu(X(t',s') - X(t,s'))\} \\
&= E\exp\{iuX(z,z']\} \cdot E\exp\{iu(X(t',s) - X(t,s))\} \\
&= E\exp\{iuX(z,z']\} \cdot \exp\left\{-\frac{u^2}{2}(t'-t)s\right\}.
\end{aligned}$$

Thus,

$$E\exp\{iuX(z,z']\} = \exp\left\{-\frac{u^2}{2}(t'-t)(s'-s)\right\},$$

which means that $X(z, z']$ is normally distributed with mean 0 and variance $(t' - t)(s' - s)$. □

Below we quote the important property of strong martingales.

Theorem 1.4 [42] *Let* $\xi(z) = \xi(t,s), (\Im_z),\ z \in [0,T]^2$ *be (with probability 1) a continuous bi-submartingale, and, moreover,* $E|\xi(z)|^p < \infty$, *for* $p > 1$. *Then the following inequality holds true:*

$$E\left(\sup_{z\in(0z']} |\xi(z)|^p\right) \le \left(\frac{p}{p-1}\right)^{2p} \sup_{z\in(0z']} E|\xi^+(z)|^p,$$

where $\xi^+(z) = \max(0, \xi(z))$

Proof Note that

$$\eta(t) := \sup_{s\in[0,T]} \xi(t,s)$$

is $\Im^1_z$-sub-martingale. Denote

$$\xi^0 := \sup_{0\le t,\, s\le T} \xi(t,s) = \sup_{0\le t\le T}\ \sup_{0\le s\le T} \xi(t,s).$$

Hence, $\xi^0 = \sup_{0\le t\le T} \eta(t)$. Therefore [44, 56],

$$E\left(\sup_{0\le t\le T} \eta(t)\right)^p \le \left(\frac{p}{p-1}\right)^p \sup_{0\le t\le T} E(\eta^+(t))^p.$$

Moreover, for fixed t we have

$$E(\eta^+(t))^p = \sup_{0\le s\le T} E(\xi^+(t,s))^p \le \left(\frac{p}{p-1}\right)^p \sup_{0\le s\le T} E(\xi^+(t,s))^p.$$

The required property follows from two inequalities above. □

Similar results for bi-martingales in discrete time are obtained in [23].

The generalization of the Burkholder inequalities can be obtained in a somewhat more complicated way, using several times the properties of one-parameter martingales. This generalization is proved by Ch. Metraux [53] for martingales under the commutation condition, but it is true for bi-martingales without the Cairoli–Walsh condition.

Let $\xi = \{\xi_{kl}\}$, $(k,l) \in N_+^2$ be a bi-martingale and put $\Delta\xi_{kl} = \xi_{k+1,l+1} - \xi_{k+1,l} - \xi_{k,l+1} + \xi_{kl}$.

Theorem 1.5 [23] *Let $\{\xi_{kl}\}$ be a bi-martingale, $\xi_{0n} = \xi_{m0} = 0$ for any n, $m \ge 0$ and $p > 1$. There exist constants c_p' and c_p'', independent of the bi-martingale $\{\xi_{kl}\}$, such that*

$$c_p' E\sigma_{mn}^p \le E|\xi_{mn}|^p \le c_p'' E\sigma_{mn}^p,$$

where

$$\sigma_{mn}^2 = \sum_{k,\,l=1}^{m-1,\,n-1} (\Delta\xi_{kl})^2.$$

Further we give the generalization of the Burkholder inequality for two-parameter martingales with discrete parameters.

Definition 1.6 The two-index sequence $\zeta = (\zeta_{kl})$, $(k,l) \in N_+^2$ given by

$$\zeta_{mn} = \sum_{k,\,l=1}^{m-1,\,n-1} \nu_{kl}\Delta\xi_{kl},$$

with the matrix (ν_{kl}), $(k,l) \in N_+^2$ consisting of $\Im_{kl}$-measurable elements such that $\sup_{k,l}|\nu_{kl}| \leq 1$ is called the Burkholder transformation $\zeta = \nu \cdot \xi$ of a bi-martingale ξ.

Burkholder transformation transforms a bi-martingale into a bi-martingale.

Theorem 1.6 [23] *For any $p > 1$ there exists a constant c_p, independent of the bi-martingale $\{\xi_{kl}\}$ and of (ν_{kl}), such that*

$$E\left[\sup_{k,l}|\zeta_{kl}|^p\right] \leq c_p \sup_{k,l} E|\xi_{kl}|^p.$$

Some new ideas for estimation of martingales were proposed by J. Brossard and L. Chevalier [6]. They use the commutation condition in two-dimensional case, and show that if the martingale ξ vanishes on the coordinate axes, the following inequality takes place:

$$P\left\{\sup_{n,m}|\xi_{mn}| > N\right\} \leq c\left(P\{\varsigma > N\} + \frac{1}{N^2}E\left[\varsigma^2\chi_{\{\varsigma\leq N\}}\right]\right),$$

where $\varsigma^2 = \sum_{m,n} E\left[(\Delta\xi_{mn})^2/\Im_{mn}\right]$, *and c is a universal constant. Hence, if $p \in (0, 2]$, one has*

$$E\left[\sup_{k,l}|\xi_{kl}|^p\right] \leq c_p E\varsigma^p.$$

Hence, the variable $\sup|\xi_{mn}|$ is finite almost surely, and the limits $\xi_{m\infty}$ and $\xi_{\infty n}$ exist almost surely on the set $\{\varsigma < \infty\}$.

The important concept in the theory of random fields is the notion of a stopping time and a stopping area.

Definition 1.7 A random variable ζ on $R_+^2 \cup \{\infty\}$ is called the stopping time if $\{\zeta \leq z\} \in \Im_z$ for any z.

However, the concept of a stopping time is not sufficient for the investigation of many problems in two-argument random fields theory. For example, if $\zeta = (\sigma,\tau)$ is a stopping time, and ξ is a progressively measurable field, then the field $\xi(\sigma \wedge t, \tau \wedge s)$ is, generally speaking, not adapted. Therefore, in the two-dimensional situation one needs the notion of the stopping line.

Definition 1.8 Let $A = A(\omega)$ be a random set. Denote the envelope of A as an open random set $(A,\infty)(\omega) = \bigcup_{z\in A(\omega)} (z,\infty)$, and its closure $[A,\infty)$ as the closed envelope of A. The set $D_A = [A, \infty)\backslash(A,\infty)$ is called the debut of A. It is progressive, if A is progressive. The set $B(\omega)$ is called the stopping line, if it is a debut of the progressive set A.

For instance, if L is a stopping line, then if $D := (-\infty, L]$, $\chi_D(u)$ is the indicator function, and the "killed" marginal is given by the formula $\mu^L(z) = \int_{[0,z]} \chi_D(u)\mu(du)$, then $\mu^L(z)$ is not constant outside D.

In some cases the more general concept, namely, the stopping area, is used.

Definition 1.9 [58, 59] The set $A \subseteq D_T \times \Omega$ is called the stopping area, if the random field $\chi_A(z)$ is $\mathfrak{I}_z$ -adapted, for any ω the cut $A(\omega)$ is closed, and $z \in A(\omega)$ implies $D_z \subset A(\omega)$.

In what follows we mostly consider the square integrable martingales.

Definition 1.10 A random function $\varphi(z)$, $z \in R_+^2$ is called increasing, if it is $(\mathfrak{I}_z)$-adapted,

$$\varphi(t,0) = \varphi(0,s) = 0, t \geq 0, s \geq 0$$

and $\varphi(z, z'] \geq 0$ for any $z, 0 \leq z \leq z'$.

Following Gikhman [23] we introduce the next definitions.

Definition 1.11 The weak characteristic of a continuous square integrable martingale is a continuous increasing function $\gamma(z) = \gamma(t,s)$ such that for any $z, z' \in R_{+}^2$, $z < z'$ the equality

$$E\left[\left(\xi(z,z']\right)^2/\mathfrak{I}_z\right] = E\left[\gamma(z,z']/\mathfrak{I}_z\right]$$

holds true.

Definition 1.12 A weak characteristic is called the characteristic (or the strong characteristic), if

$$E\left[\left(\xi(z,z']\right)^2/\mathfrak{I}_z^*\right] = E\left[\gamma(z,z']/\mathfrak{I}_z^*\right]$$

for any $z, z' \in R_{+}^2$, $z < z'$.

We quote without a proof the following statements [23].

Theorem 1.7 *If $\xi(z)$ is a square integrable bi-martingale and $\gamma(z)$ is its weak characteristic, then*

$$E\left[\left(\xi(z,z']\right)^2/\mathfrak{I}_z\right] = E\left[\gamma(z,z']/\mathfrak{I}_z\right],$$

which means that the random process $\xi^2(z) - \gamma(z)$ is a weak martingale.

Theorem 1.8 *If $\xi(z)$ is a square integrable strong martingale and $\gamma(z)$ is its characteristic, then the process $\xi^2(z) - \gamma(z)$ is a bi-martingale.*

Theorem 1.9 [23] *For a square integrable bi-martingale $\xi(z)$ there exist functions $\alpha(z)$ and $\beta(z)$ with the following properties:*

(a) $\alpha(z)$ and $\beta(z)$ are $\mathfrak{I}_z$ -adapted,
(b) $\alpha(z)$ is monotone non-decreasing under fixed s, and continuous in s, $\alpha(0,s) = 0$,
(c) $\beta(z)$ is monotone non-decreasing under fixed t, and continuous in t, $\beta(t,0) = 0$,
(d) $\alpha(z)$ is the characteristic of the martingale $\left(\xi_{\cdot s}(t), \mathfrak{I}_z^1,\ \ t \geq 0\right)$, $\beta(z)$ is the characteristic of the martingale $\left(\xi_{t\cdot}(s), \mathfrak{I}_z^2,\ \ s \geq 0\right)$,
(e) for fixed t and Δt ($t \geq 0$, $\Delta t > 0$), the function $\Delta_t\alpha(t,s) = \alpha(t + \Delta t,\ s) - \alpha(t, s)$ is a $\mathfrak{I}_{(t,s)}$ -sub-martingale,
(f) for fixed s and Δs ($s \geq 0$, $\Delta s > 0$), the function $\Delta_s\beta(t,s) = \beta(t,\ s + \Delta s) - \beta(t, s)$ is a $\mathfrak{I}_{(t,s)}$ - sub-martingale.

If $\xi(z)$ is a strong martingale, then

$$E\left[\left(\xi\left(z,z'\right]\right)^2/\mathfrak{I}_z^1\right] = E\left[\alpha\left(z,z'\right]/\mathfrak{I}_z^1\right]$$

and

$$E\left[\left(\xi\left(z,z'\right]\right)^2/\mathfrak{I}_z^2\right] = E\left[\beta\left(z,z'\right]/\mathfrak{I}_z^2\right].$$

Proof Take $0 < z < z'$. Putting $(0, z'] = (0, z] \cup A \cup B \cup (z, z']$, where A and B are rectangles, we easily obtain that

$$\begin{aligned}\xi\eta\left(z,z'\right] = {} & \left(\xi(A) + \xi(B) + \xi\left(0,z\right]\right)\eta\left(z,z'\right] + \left(\eta(A) + \eta(B) + \eta\left(0,z\right]\right)\xi\left(z,z'\right] \\ & + \xi(A)\eta(B) + \xi(B)\eta(A) + \xi\left(z,z'\right]\eta\left(z,z'\right].\end{aligned}$$

Hence, if ξ and are square-integrable bi-martingales, then

$$E\left[\xi\eta\left(z,z'\right]/\mathfrak{I}_z\right] = E\left[\xi\left(z,z'\right]\eta\left(z,z'\right]/\mathfrak{I}_z\right]. \tag{1.2}$$

If ξ and η are square-integrable strong martingales, then

$$\begin{aligned}E\left[\xi\eta\left(z,z'\right]/\mathfrak{I}_z^1\right] &= E\left[\xi\left(z,z'\right]\eta\left(z,z'\right]/\mathfrak{I}_z^1\right], \\ E\left[\xi\eta\left(z,z'\right]/\mathfrak{I}_z^2\right] &= E\left[\xi\left(z,z'\right]\eta\left(z,z'\right]/\mathfrak{I}_z^2\right]\end{aligned} \tag{1.3}$$

Observe that (1.2) implies that if ξ is a square-integrable bi-martingale, and $\gamma(z)$ is its weak characteristic, then

$$E\left[\xi^2\left(z,z'\right]/\mathfrak{I}_z\right] = E\left[\gamma\left(z,z'\right]/\mathfrak{I}_z\right],$$

that is, the field $\xi^2(z) - \gamma(z)$ is a weak martingale. If ξ is a square integrable strong martingale and $\gamma(z)$ is its characteristic, then (1.3) implies immediately

$$E\left[\xi^2(z,z')/\Im_z^1\right] = E\left[\gamma(z,z')/\Im_z^1\right],$$
$$E\left[\xi^2(z,z')/\Im_z^2\right] = E\left[\gamma(z,z')/\Im_z^2\right],$$

therefore, field $\xi^2(z) - \gamma(z)$ is a bi-martingale.

Let ξ be a square integrable continuous bi-martingale with $\sup_{(t,s)} E\xi^2(z) < \infty$. Then for fixed s the process $\xi(t,s)$ is a continuous square integrable martingale with respect to both flows ($\Im_z, t \geq 0$, s is fixed) and $\left(\Im_z^1, t \geq 0\right)$.

The existence of the one-parametric martingale characteristic implies the existence of such increasing continuous in t under fixed s process $\alpha(t,s) = \alpha(z)$, adapted to the flow $\left(\Im_z^1, t \geq 0\right)$, that $\xi^2(t,s) - \alpha(t,s)$ is a $\left(\Im_z^1\right)$ -martingale. Thus,

$$E\left[\xi((t+\Delta t,s) - \xi(t,s))/\Im_z^1\right] = E\left[(\alpha(t+\Delta t,s) - \alpha(t,s))/\Im_z^1\right].$$

The continuity of the one-parameter process $\xi_s(t) = \xi(t,s)$ under fixed s implies that $\alpha_s(t) = \alpha(t,s)$ almost surely coincides with the square variance of the field slice $\xi_s(t)$ on the segment $[0,t]$. Therefore, $\alpha_s(t)$ is $\Im_z$ -measurable. Thus, the function $\alpha_s(t)$ is adapted to the flow $(\Im_z, t \geq 0)$, and the process $\xi_s(t) - \alpha_s(t)$ is a $(\Im_z, t \geq 0)$ -martingale. Then (1.2) implies

$$E\left[\xi^2(z,z')/\Im_z\right] = E\left[\alpha(z,z')/\Im_z\right]. \tag{1.4}$$

The function $\alpha(z)$, $z = (t,s)$, is a monotone non-decreasing function of argument t under fixed s, but, generally speaking, it is not the increasing function of two arguments. On the other hand, equality (1.4) shows that the difference $\Delta_t\alpha_{t\cdot}(s) = \alpha(t + \Delta t, s) - \alpha(t,s)$ is $(\Im_z, s \geq 0)$ -sub-martingale (as a function of s for any $\Delta t > 0$, $s \geq 0$). Thus, for $\alpha(z)$ the statements (a), (b), (d), (e) hold true.

If, moreover, $\xi(z)$ is a strong martingale, then $E\left[\xi^2(z,z')/\Im_{z_i}^1\right] = E\left[\alpha(z,z')/\Im_{z_i}^1\right]$. The existence and the properties of function $\beta(z)$ can be proved similarly. □

Definition 1.13 Functions $\alpha(z)$ and $\beta(z)$ are called the semi-characteristics of the bi-martingale $\xi(z)$.

Sometimes we will also use the notation $< \xi, \xi >_z$ for the characteristic of the martingale $\xi(z)$.

Remark 1.9 If the characteristic $\gamma(z)$ of a strong martingale exists, it can be easily related with the functions $\alpha_{\cdot s}(t)$ and $\beta_{t\cdot}(s)$, which were constructed in the proof of Theorem 1.9. Namely,

$$\alpha(z) = \gamma(z) + \alpha_0(t), \beta(z) = \gamma(z) + \beta_0(s),$$

where $\alpha_0(t)$ is the characteristic of the martingale ($\xi(t,0)$, $t \geq 0$), and $\beta_0(s)$ is the characteristic of the martingale ($\xi(0,s)$, $s \geq 0$). This representation implies the uniqueness of the characteristic of a strong martingale.

Below we provide the conditions for the existence of the characteristic of a strong martingale. For this conditions we refer to [21, 23] and follow these papers in our presentation below.

Take a rectangle $D_{ab} = [0,a] \times [0,b]$ and divide it using straight lines parallel to the coordinate axes passing through the points z_{ij}, $0 = t_0 < t_1 < \ldots < t_{n+1} = a$ and $0 = s_0 < s_1 < \ldots < s_{m+1} = b$ (we denote this partition by λ). Further, consider some strong martingale $\mu(z) \in sM_2^c$ and put $S_\lambda = \sum_{i,j=0}^{n,m} \left(\Delta\mu_{ij}\right)^2$, where $\Delta\mu_{ij} = \mu(z_{ij}, z_{i+1,j+1}]$.

Lemma 1.1 *The family of random variables* $\{S_\lambda\}$ *is uniformly integrable.*

Proof Introduce the sequence of random variables:

$$\xi_0 = E\left\{(\mu_{a\cdot}(b))^2/\Im_{(a,0)}\right\},$$

$$\xi_j = E\left\{(\mu_{a\cdot}(b))^2/\Im_{(a,0)}\right\} - \left(\mu_a(s_j)\right)^2 + \sum_{i=0}^{n} \left(\Delta\mu_{ij}\right)^2, j = \overline{1, m+1}.$$

We prove that the sequence $\left\{\xi_j, j = \overline{0, m+1}\right\}$ is a nonnegative supermartingale. Note that

$$\mu\left(a, \Delta y_{j-1}\right) = \mu\left(a, y_j\right) - \mu\left(a, y_{j-1}\right) = \sum_{i=0}^{n} \Delta\mu_{i,j-1}$$

and, therefore,

$$E\left\{\left(\mu\left(a, \Delta y_{j-1}\right)\right)^2/\Im_{\left(a,s_{j-1}\right)}\right\} = E\left\{\left(\sum_{i=0}^{n} \Delta\mu_{i,j-1}\right)^2/\Im_{\left(a,s_{j-1}\right)}\right\}.$$

If $i < k$, then

$$E\left\{\Delta\mu_{i,j-1}\Delta\mu_{k,j-1}/\Im_{\left(a,s_{j-1}\right)}\right\} = E\left\{\Delta\mu_{i,j-1}E\left\{\Delta\mu_{k,j-1}/\Im^*_{\left(t_k,s_{j-1}\right)}\right\}/\Im_{\left(a,s_{j-1}\right)}\right\} = 0.$$

Thus,

$$E\left\{\left(\mu\left(a, \Delta y_{j-1}\right)\right)^2/\Im_{\left(a,s_{j-1}\right)}\right\} = E\left\{\sum_{i=0}^{n} \left(\Delta\mu_{i,j-1}\right)^2/\Im_{\left(a,s_{j-1}\right)}\right\}.$$

Hence,

$$E\left\{\xi_j/\mathfrak{I}_{(a,s_{j-1})}\right\} = E\left\{(\mu_{a\cdot}(b))^2/\mathfrak{I}_{(a,s_{j-1})}\right\} - (\mu_{a\cdot}(s_{i-1}))^2 \le \xi_{j-1}.$$

Thus, $\left\{\xi_j, \mathfrak{I}_{(a,s_j)},\ j = 0, \ldots, \ m+1\right\}$ is a super-martingale. The last equality implies also that $\xi_{j-1} \ge 0, j = \overline{1, m+1}$. Moreover,

$$\xi_{m+1} = \sum_{i=0}^{n} (\Delta\mu_{im})^2 \ge 0.$$

Introduce the monotone increasing sequence $\alpha_j, j = \overline{0, m+1}, \alpha_0 = 0$, associated with the super-martingale ξ_j, and put

$$\Delta\alpha_j = \alpha_{j+1} - \alpha_j = -E\{\xi_{j+1} - \xi_j/\mathfrak{I}_{(a,s)}\} = \sum_{i=0}^{n} (\Delta\mu_{ij})^2,$$

$$\alpha_{m+1} = \sum_{j=0}^{m}\sum_{i=0}^{n} (\Delta\mu_{ij})^2 = S_\lambda.$$

Further, consider the sub-martingale ξ_j extended up to $j = m + 2, \ldots$, as $\xi_{m+2} = \xi_{m+3} = \ldots = 0, \mathfrak{I}_{(a,s_{m+k})} = \mathfrak{I}_{(a,s_{m+1})}, k = 1, 2, \ldots$

For an arbitrary $\mathfrak{I}_{(a,s_j)}$ -stopping time σ, taking values $1, \ldots, m + 2$, set $\nu(t) = \nu_\sigma(t) = \mu(t,s_\sigma) - \mu(t,s_{\sigma-1}), \nu(t_i) - \nu(t_{i-1}) = \Delta\mu(t_{i-1},s_{\sigma-1})$,
and let

$$\overline{\xi}_i = E\{\nu^2(a)/\mathfrak{I}_{i\sigma}\} - \nu^2(t_i) + (\nu(t_i) - \nu(t_{i-1}))^2.$$

It is easy to see that $\overline{\xi}_i$ is a nonnegative super-martingale, and for the sequence $\overline{\alpha}_i$, associated with $\overline{\xi}_i$, the equalities

$$\Delta\overline{\alpha}_i = \overline{\alpha}_i - \overline{\alpha}_{i-1} = (\nu(t_i) - \nu(t_{i-1}))^2,$$

$$\overline{\alpha}_{n+1} = \sum_{i=0}^{n} (\Delta\mu(t_{i-1}, s_{\sigma-1}))^2$$

take place.

Let τ be an arbitrary stopping time with respect to the flow $\{\mathfrak{I}_{i\sigma},\ i = 0, 1, \ldots,\ n+1\}$. Then

$$P\{|\overline{\xi}_\tau| > c\} = \frac{1}{c}E\left\{(\mu(a, y_\sigma) - \mu(a, s_{\sigma-1}))^2 + 4\sup|\mu(z)|^2\right\}$$

(here by sup we mean taking the lid of the rectangle D_{ab}), implying $P\{|\overline{\xi}_\tau| > c\} \to 0$ as $c \to \infty$, uniformly in σ, τ, and λ. Moreover,

$$\int_{\{\overline{\xi}_\tau > c\}} \overline{\xi}_\tau dP \le \int_{\{\overline{\xi}_\tau > c\}} E\left\{(\mu(a,s_\sigma) - \mu(a,s_{\sigma-1}))^2/\mathfrak{I}_{\tau\sigma}\right\}dP + \int_{\{\overline{\xi}_\tau > c\}} 4\sup\mu^2(z)dP.$$

In the last expression, the second integral converges uniformly to 0 as $c \to \infty$ because the variable sup $\mu^2(z)$ is integrable, and $P\{|\overline{\xi}_\tau| > c\} \to 0$ uniformly. Since $\{|\overline{\xi}_\tau| > c\}$ is measurable,

$$\int_{\{\overline{\xi}_\tau > c\}} E\left\{(\mu(a,s_\sigma) - \mu(a,s_{\sigma-1}))^2/\mathfrak{I}_{\tau\sigma}\right\}dP = \int_{\{\overline{\xi}_\tau > c\}} \mu^2(a,\Delta s_{\sigma-1})dP \le \int_{\{\overline{\xi}_\tau > c\}} 4\sup\mu^2(a,s)dP$$

and converges to 0 uniformly. Therefore,

$$\sup_{\tau,\sigma,\lambda} P\{|\overline{\xi}_\tau| > c\} \to 0 \text{ as } c \to \infty.$$

This implies, in particular [27], that the set of variables $\overline{\alpha}_{n+1} = \alpha_{n+1}(\tau,\sigma,\lambda)$ is uniformly integrable. Thus, $\sum_{i=0}^{n} (\Delta\mu(t_{i-1}, s_{\sigma-1}))^2$ is uniformly integrable.

Let us return to the super-martingale ξ_j introduced above. For ξ_j we have

$$E\left\{\xi_j/\mathfrak{I}_{(a,s_{j-1})}\right\} = E\left\{\mu_a^2(b)/\mathfrak{I}_{(a,s_j)}\right\} + \sum_{i=0}^{n} \left(\Delta\mu\left(t_{i-1}, s_{j-1}\right)\right)^2.$$

Now let σ be a random time moment $\left\{\mathfrak{I}_{(a,s_j)},\ j = 0, \ldots, \ m+1\right\}$. Then

$$P\{\xi_\sigma > c\} \le \frac{1}{c}\left(E\mu^2(a,b) + E\sum_{i=0}^{n} (\Delta\mu(t_{i-1}, s_{\sigma-1}))^2\right).$$

Due to the uniform integrability of $\overline{\alpha}_{n+1}$ the variables $E\overline{\alpha}_{n+1}$ are uniformly bounded. Hence $P\{\xi_\sigma > c\}$ vanishes uniformly in λ and σ. Moreover,

$$\int_{\{\xi_\sigma > c\}} \xi_\sigma dP \le \int_{\{\xi_\sigma > c\}} E\left\{\mu_a^2(b)/\mathfrak{I}_{(a,s_j)}\right\}dP + \int_{\{\xi_\sigma > c\}} \sum_{i=0}^{n} \left(\Delta\mu\left(t_{i-1}, s_{j-1}\right)\right)^2 dP$$

$$= \int_{\{\xi_\sigma > c\}} \mu^2(a,b)dP + \int_{\{\xi_\sigma > c\}} \sum_{i=0}^{n} \left(\Delta\mu\left(t_{i-1}, s_{j-1}\right)\right)^2 dP$$

Since $P\{\xi_\tau > c\} \to 0$ uniformly in λ and σ, and the sums $\sum_{i=0}^{n} (\Delta\mu_{i\sigma})^2$ are uniformly integrable, both integrals in the right-hand side of the inequality tend to 0, uniformly in λ and σ, which implies that the set $\{S_\lambda\}$ of random variables is uniformly integrable. □

Lemma 1.2 *$P \cdot \lim_{|\lambda|\to 0} S_\lambda$ exists, and*

$$P \cdot \lim_{|\lambda|\to 0} S_\lambda = P \cdot \lim_{|\lambda|\to 0} \sum_{i=0}^{n}\sum_{j=0}^{m} (\Delta\mu_{ij})^2 = \alpha_a(b) - \alpha_0(b).$$

Proof For $0 \le t \le a$ put

$$\mu(t, \Delta s_j) = \mu(t, s_{j+1}) - \mu(t, s_j),$$

$$\overline{\sigma}_m^2 = \sum_{j=0}^{m} (\mu(a, \Delta s_j))^2.$$

As proved above, the sums $\overline{\sigma}_m^2 = \overline{\sigma}^2(\lambda)$ are uniformly integrable and converge to $\alpha_a(b)$ in probability as $|\lambda| \to 0$. Note that

$$\mu(a, \Delta s_j) - \mu(a, \Delta s_{j-1}) = \mu((0, s), (a, s_{j+1})] = \sum_{i=0}^{n} \Delta\mu_{ij},$$

implying

$$\alpha_a(b) = P \cdot \lim_{|\lambda|\to 0} \overline{\sigma}_m^2 = P \cdot \lim_{|\lambda|\to 0} (A + B + 2C),$$

where

$$A = \sum_{j=0}^{m} (\mu(0, \Delta s_j))^2, B = \sum_{j==0}^{m} \left(\sum_{i=0}^{n} \Delta\mu_{ij}\right)^2,$$

$$C = \sum_{j=0}^{m} \mu(0, \Delta s_j)\mu((0, s), (a, s_{j+1})].$$

The expression A is uniformly (with respect to λ) integrable, and $P \cdot \lim_{|\lambda|\to 0} A = \alpha_0(b)$, where $\alpha_0(b)$ is the characteristic of the martingale $\mu(0,s)$.

Let us prove that C converges to 0 in probability as $|\lambda| \to 0$. Define the stopping time as $\tau = \inf\{x : |\mu(0,x) \ge N|, x \in [0,b]\}$, if the set in brackets is nonempty, and $\tau = b$ otherwise. Then $\mu'(s) = \mu(0, s \wedge \tau)$ is a $\mathfrak{I}_{(0,s)}$ -martingale, and $|\mu'(s)| \le N$. Put

$$\eta_k = \sum_{j=0}^{k} \mu'(\Delta s_{j-1})\mu((0, s), (a, s_{j+1})], \qquad k = 1, \ldots, m + 1.$$

We have $P\{C \ne \eta_{m+1}\} \le P\{\tau < b\} \to 0$ as $N \to \infty$. All terms in η_k have finite second moments, and their orthogonality can be checked in the same fashion as above. Hence,

$$E\eta_{m+1}^2 = E\sum_{j=1}^{m+1} \left(\mu'(\Delta s_{j-1})\right)^2 \left(\mu\left((0,s_{j-1}),(a,s_j)\right]\right)^2.$$

Note that $S' = \sum_{j=1}^{m+1} \left(\mu'(\Delta s_{j-1})\right)^2 \left(\mu\left((0,s_{j-1}),(a,s_j)\right]\right)^2$ is uniformly integrable.

Indeed,

$$S' \le 4N^2 \sum_{j=1}^{m+1} \left(\left(\mu'(\Delta s_{j-1})\right)^2 + \mu^2(a,s_{j-1})\right),$$

and the right-hand side is uniformly integrable, which is the well-known result for one-parametric martingales, see [26]. Moreover,

$S' \le \max_j \left|\mu'(\Delta s_{j-1})\right|^2 \sum_{j=1}^{m+1} \left(\mu^2(0,\Delta s_{j-1}) + \mu^2(a,\Delta s_{j-1})\right) \to 0$ as $|\lambda| \to 0$, that is, S' converges to 0 with probability 1.

From the upper estimate

$$P\{|C| > \varepsilon\} \le P\{C \ne \eta_{m+1}\} + P\{|\eta_{m+1}| > \varepsilon\} \le P\{\tau < b\} + \frac{E\eta_{m+1}^2}{\varepsilon^2}$$

we immediately derive

$$P \cdot \lim_{|\lambda|\to 0} C = 0.$$

Consider now the sum B. Since $B = S_\lambda + 2B_2$, $B_2 = \sum_{j=0}^{m}\sum_{i=0}^{n} \Delta\mu_{i+1,j}\,\mu\left((0,s_j),\right.$ $\left.(t_i,s_{j+1})\right]$, it remains to show that $P \cdot \lim_{|\lambda|\to 0} B_2 = 0$. We have

$$\mu\left((0,s_j),(t_i,s_{j+1})\right] = \xi_{j+1}(t_i) - \xi_{j+1}(0),$$

where $\xi_j(t) = \mu(t,s_j) - \mu(t,s_{j-1})$ is a sequence of continuous square integrable $\mathfrak{I}_z^1$-martingales.

Put

$$\tau_j = \inf\left\{t : \left|\xi_j(t)\right| \ge N\right\}, (\inf \varnothing = a),$$

$$\xi'_j(t) = \xi_j(t \wedge \tau_j),$$

$$B'_2 = \sum_{j=0}^{m}\sum_{i=0}^{n} \left(\xi'_j(t_i) - \xi'_j(0)\right)\Delta\mu_{i+1,j}.$$

For $j < k$

$$E\left(\xi'_j(t_i) - \xi'_j(0)\right)\Delta\mu_{i+1,j}\left(\xi'_k(t_l) - \xi'_k(0)\right)\Delta\mu_{l+1,k}$$
$$= E\left[\ldots E\left[\Delta\mu_{l+1,k}/\mathfrak{I}^1_{(t_{l+1},s_k)}\right]\right] = 0.$$

Thus,

$$E\left(B'_2\right)^2 = E\sum_{j=0}^{m}\left(\sum_{i=0}^{n}\left(\xi'_j(t_i) - \xi'_j(0)\right)\Delta\mu_{i+1,j}\right)^2 = EB_3 + EB_4,$$

$$B_3 = \sum_{j=0}^{m}\sum_{i=0}^{n}\left(\xi'_j(t_i) - \xi'_j(0)\right)^2(\Delta\mu_{i+1,j})^2,$$

$$B_4 = 2\sum_{j=0}^{m}\sum_{i<k}\left(\xi'_j(t_i) - \xi'_j(0)\right)\Delta\mu_{i+1,j}\left(\xi'_j(t_k) - \xi'_j(0)\right)\Delta\mu_{k+1,j}.$$

It is obvious that the mean of each summand in the sum B_4 is 0. For B_3 we derive

$$|B_3| \le 2N^2\sum_{j=0}^{m}\sum_{i=0}^{n-1}(\Delta\mu_{i+1,j})^2,$$

which implies its uniform integrability. On the other hand,

$\max_{i,j}\left(\xi'_j(t_i) - \xi'_j(0)\right) \to 0$ as $|\lambda| \to 0$

with probability 1. Therefore, $EB_3 \to 0$ as $|\lambda| \to 0$ and, consequently, $E(B'_2)^2 \to 0$. Further,

$$P\left\{B_2 \ne B'_2\right\} \le P\left\{\bigcup_j\left\{\max\left|\xi'_j(t) - \xi_j(t)\right| > 0\right\}\right\} \le P\left\{\max_z\left|\mu(z) > \frac{N}{2}\right|\right\} \to 0$$

as $N \to \infty$

uniformly in λ. If we first choose N such that $P\{B_2 \ne B'_2\} \le \varepsilon/2$ for all λ, and then λ_0 such that $P\{|B'_2| > \varepsilon\} < \varepsilon$ for λ: $|\lambda| < |\lambda_0|$, we obtain

$P\{|B_2| > \varepsilon\} < \varepsilon$ for all $|\lambda| < |\lambda_0|$.

The existence of $P\cdot\lim_{|\lambda|\to 0} S_\lambda = \alpha_a(b) - \alpha_0(b)$ follows from the equality

$$P\cdot\lim_{|\lambda|\to 0} B = \alpha_a(b) - \alpha_0(b).$$ □

Definition 1.13 The limit $P\cdot\lim_{|\lambda|\to 0} S_\lambda$ is called the square variation of the function $\mu(z)$ on the rectangle D_{ab}.

Put $\gamma(a,b) = P \cdot \lim_{|\lambda| \to \infty} S_\lambda$. The properties below follow straightforwardly from the definition of $\gamma(z)$:

(a) $\gamma(z) \geq 0$,
(b) $\gamma(0,s) = \gamma(t,0) = 0$,
(c) $\gamma(z)$ is adapted to the flow $\{\Im_z, z \in R_+^2\}$,
(d) $\gamma(0,z] \geq 0$.

Theorem 1.10 *For any strong square integrable martingale* $\xi(z)$ *there exists a unique continuous characteristic.*

Proof Note that

$$E\left[\left(\mu(z,z']\right)^2/\Im_z^*\right] = E\left[\left(\sum_{i,j=0}^{n,m} \Delta\mu_{ij}\right)^2 \Big/ \Im_z^*\right] = E\left[\sum_{i,j=0}^{n,m} (\Delta\mu_{ij})^2 \Big/ \Im_z^*\right],$$

where $z_{ij} = (t_i, s_j)$, $t = t_0 < t_1 < \ldots < t_{n+1} = t'$ and $s = s_0 < s_1 < \ldots < s_{m+1} = s'$, $\Delta\mu_{ij} = \mu(z_{ij}, z_{i+1,j+1}]$.

Since the sum $\sum_{i,j=0}^{n,m} (\Delta\mu_{ij})^2$ is uniformly integrable, we have $P.\lim_{|\lambda| \to \infty} \sum_{i,j=0}^{n,m} (\Delta\mu_{ij})^2 = \gamma(z,z']$, and passing to the limit as $|\lambda| \to 0$, we obtain

$$E\left[\left(\mu(z,z']\right)^2/\Im_z^*\right] = E\left[\left(\gamma(z,z']\right)/\Im_z^*\right]. \qquad \square$$

For continuous bi-martingales a weaker version of the previous theorem holds true.

Theorem 1.11 *Let* $\gamma(z)$ *be a weak characteristic of martingale* $\mu \in bM_2^c$. *Then*

$$\gamma(z) = \lim_m \lim_n \sum_{i=0}^{m} \sum_{j=0}^{n} (\Delta\mu_{ij})^2 = \lim_n \lim_m \sum_{j=0}^{n} \sum_{i=0}^{m} (\Delta\mu_{ij})^2.$$

The proof is analogous to that of Theorem 1.8.

1.2 Stochastic Two-Parametric Integrals

Theory of stochastic integrals with respect to bi-martingales is constructed partly in the same way as in the one-dimensional case. In future exposition we confine ourselves to the integration with respect to continuous square integrable bi-martingales.

Let $(\xi(z), \Im_z)$, $z \in R_+^2$, be a continuous square integrable bi-martingale. Denote by $\boldsymbol{B}_0$ the class of functions, which can be represented in the form

$$\eta(z) = \sum_{k=1}^{n} \eta_k \chi_{(z_k, z'_k]},$$

where $z_k < z'_k$, $(z_k, z'_k]$ are disjoint rectangles, η_k are bounded $\Im_{z_k}$-measurable random variables, $k = \overline{1,n}$, and $\chi_A(z)$ is the indicator of a set A.

Definition 1.16 A continuous increasing function $\gamma(z)$ is called a weak characteristic of bi-martingale $\xi(z)$ if for any $\{z,z'\} \in R_+^2$, $z < z'$, we have

$$E\left\{\left(\xi(z,z']\right)^2/\Im_z\right\} = E\left\{\lambda(z,z']/\Im_z\right\}.$$

Definition 1.17 For the class of functions from $\boldsymbol{B}_0$ define a stochastic integral $I(\eta)$ as

$$I(\eta) := \int_{R_+^2} \eta(z)\xi(dz) = \sum_{k=1}^{n} \eta_k \xi(z_k, z'_k].$$

A stochastic integral can be determined on the closure in L_2 of the class $\boldsymbol{B}_2(\Im_z, \gamma)$ of random functions $\eta = \eta(z)$ which have the following property: there exists a sequence $\eta_n \in \boldsymbol{B}_0$ such that

$$\lim_{n\to\infty} E\int_{R_+^2} (\eta(z) - \eta_n(z))^2 \gamma(dz) = 0,$$

where $\gamma(z)$ is the weak characteristic of $\xi(z)$.

Determined stochastic integral $I(z,\eta) = \int_{[0,z]} \eta(u)\xi(du)$ is determined by the equality:

$$I(z,\eta) = I\left(\chi_{[0,z]}(\eta)\right).$$

Theorem 1.12 *Let* I(z,η) *be a stochastic integral. Then:*

(a) $I(z,\eta)$, $\eta \in \boldsymbol{B}_2(\Im_z, \gamma)$ is the square integrable bi-martingale with weak characteristic $[I,I](z,\eta) = \int_{[0,z]} \eta^2(u)\gamma(du)$;
(b) the field $I(z,\eta)$, $z \geq 0$, has a continuous modification;
(c) if $\xi(z)$ is a square integrable strong martingale, then $I(z,\eta)$ is a square integrable strong martingale as well;

(d) if $\hat{\eta}, \tilde{\eta} \in \boldsymbol{B}_2(\Im_z, \gamma)$ and $\hat{\eta}(z,\omega) = \tilde{\eta}(z,\omega)$ for any $z \in R_+^2$ and $\omega \in \Lambda$, then $I(z\hat{\eta}) = I(z\tilde{\eta})$ almost surely for any $z \in R_+^2$ and $\omega \in \Lambda$;

(e) if $\xi(z)$ is a strong martingale, then the inequality

$$P(|I(z,\eta)| > \varepsilon) \leq \frac{\delta}{\varepsilon^2} + P\left(\int\limits_{[0,z]} \eta^2(u)\gamma(dy) > \delta\right)$$

holds true for any $\delta > 0$, $\varepsilon > 0$.

The proof of these properties is similar to the proof in the one-dimensional case, and is given, for instance, in [19, 23, 24].

Definition 1.18 A random field $\xi(z) = \xi(z,\omega)$, $z \in R_+^2$, with values in the phase space $(X, \mathscr{B})$ is called progressively measurable if for any z the function $\xi(z',\omega)$, $z' < z$, is measurable on the probability space $(\Omega, \Im_z, P)$, that is, it is measurable with respect to the σ -algebra $\mathscr{B}_z \otimes \Im_z$, where $\mathscr{B}_z$ is a σ -algebra of Borel functions on [0,z].

Remark 1.10 The stochastic integral defined above can be extended to the class of progressively measurable functions $\eta(z)$, for which the integral

$$\int\limits_{[0,z]} \eta^2(u)\gamma(du)$$

exists almost surely for all $z > 0$.

Remark 1.11 For $\Im_z^1$ - and $\Im_z^2$ -measurable functions the integral $I(z)$ can be determined similarly to the integral $I(z)$ for $\Im_z$ -measurable functions.

Stochastic integrals with respect to the Wiener field introduced in [65, 71, 72] and generalized for arbitrary strong martingales (under condition F4) [7] are of special interest. Stochastic integrals with respect to strong martingales were investigated in [23–25, 42, 64].

Suppose that the characteristic $\gamma(z)$ of the strong martingale $\xi(z)$, $E\xi(z) = 0$, is absolutely continuous with respect to the Lebesgue measure, that is,

$$\gamma(z) = \int\limits_{[0,z]} \sigma^2(u)du \quad P\text{–a.s.}, \quad (1.3)$$

where $\sigma^2(z)$ is nonnegative -measurable function, Lebesgue integrable with probability 1.

Remark 1.12 If $\sigma^2(z) = 1$, then $\xi(z)$ is a standard Wiener field $W(dz)$. We will denote the integral with respect to $W(dz)$ by .

Theorem 1.13 [42, 47] *If the equality*

$$P\left(\int_{[0,z]} \psi^2(u)\sigma^2(u)du < \infty\right) = 1$$

holds true for any $\Im_z$ -measurable function ψ(z), then the stochastic integral

$$I(z,\psi) = \int_{[0,z]} \psi(u)\xi(du)$$

is determined with probability 1, and possesses a continuous modification. If, moreover,

$$\int_{[0,z]} E\psi^2(u)\sigma^2(u)du < \infty,$$

Then

(a) $I(z,\psi)$ is a strong martingale;
(b) $E(I(z,\psi)) = 0$;

(c) $E\left(\int_{[0,z]} \psi(u)\xi(du)\right)^2 = \int_{[0,z]} E\psi^2(u)\sigma^2(u)du;$

(d) $P\left\{\left|\int_{[0,z]} E\psi(u)\xi(du)\right| > B\right\} \leq \frac{A}{B^2} + P\left\{\left|\int_{[0,z]} E\psi(u)\xi(du)\right| > A\right\}, \quad A > 0.$

The stochastic integral with respect to the strong martingale ξ(z) for arbitrary area z' ≤ z ≤ z'' is defined similarly. Moreover, Theorem 1.11 holds true, as well as the following properties:

$$E\left[\int_{[z_1,z_2]} \psi(u)\xi(du)/\Im^*_{z_1}\right] = 0,$$

$$E\left[\left(\int_{[z_1,z_2]} \psi(u)\xi(du)\right)^2 \Big/ \Im^*_{z_1}\right] = E\left[\int_{[z_1,z_2]} \psi^2(u)\sigma^2(u)du/\Im^*_{z_1}\right],$$

provided that

$$\int_{[0,z]} E\psi^2(u)\sigma^2(u)du < \infty.$$

Denote by $\mathbf{B} = \mathbf{B}(\Im_z)$ *the class of* $\Im_z$ *-adapted random functions* $\alpha(z)$*, such that*

$$P\left\{ \int\limits_{[0,T]^2} |\alpha(z)|^2 dz < \infty \right\} = 1.$$

Denote by $\mathbf{B}_2 = \mathbf{B}_2(\Im_z)$ *a sub-space of space* $\mathbf{B}$, consisting of random functions satisfying the condition $E \int\limits_{[0,T]^2} |\alpha(z)|^2 dz < \infty$.

Theorem 1.14 [42] *Let* $\xi(z)$ *be a continuous square integrable strong martingale with zero boundary values* $\xi(t,0) = \xi(0,s) = 0$*. Suppose that for its characteristic (1.1) holds true, and* $\sigma^2(z) > 0$ P *-almost surely. Then there exists a Wiener field* W(z) *such that*

$$\xi(z) = \int\limits_{[0,z]} \sigma(u) W(du).$$

Proof Put $W(z) = \int\limits_{[0,z]} \frac{\xi(du)}{\sigma(u)}$. From the properties of the stochastic integral we see that the random field $W(z)$ is $\Im_z$ -measurable. Moreover, $E\left(W\left(z,z'\right]/\Im_z^*\right) = 0\,P$ -a.s., and $E\left(\left(W\left(z,z'\right]\right)^2/\Im_z^*\right) = \left(t'-t\right)\left(s'-s\right)$ P -a.s. Thus, Theorem 1.4 implies that the random field $W(z)$ is the Wiener field. □

Theorem 1.15 [42] *Assume that* $\xi(z)$ *is a strong Gaussian martingale, and conditions of Theorem 1.9 hold true. Then there exists a Wiener field* W(z) *such that*

$$\xi(z) = \int\limits_{[0,z]} \left(E|\sigma(u)|^2\right)^{1/2} W(du).$$

Proof Put $W(z) = \int\limits_{[0,z]} \frac{\xi(du)}{\left(E|\sigma(u)|^2\right)^{1/2}}$. Since the random field $\xi(z)$ is Gaussian, $W(z)$ is a strong Gaussian martingale. Moreover,

$$EW(z) = 0,$$

and

$$EW(z)W(z') = \min(t,t')\min(s,s'),$$

which implies that $W(z)$ is a Wiener field. □

If the random field $\psi(z)$ does not depend on $\xi(z)$, $E\psi^2(z) < \infty$, $z \in [0,T]^2$, and the sample function $\psi(z)$ has continuous mixed derivative $\frac{\partial^2\psi}{\partial t\partial s}$ with probability 1, then the integral with respect to the strong martingale $\xi(z)$ with zero boundary values can be rewritten using integration by parts:

$$I(T,\psi) = \int\limits_{[0,T]^2} \psi(u)\xi(du)$$
$$= \int\limits_{[0,T]^2} \frac{\partial^2\psi(t,s)}{\partial t \partial s}\xi(t,s)dtds \int\limits_0 \frac{\partial^2\psi(t,T)}{\partial t \partial s}\xi(t,T)dt - \int\limits_0^T \frac{\partial\psi(T,s)}{\partial s}\xi(T,s)ds$$
$$= \psi(T,T)\xi(T,T).$$

The validity of this representation follows from the sequence of equalities below:

$$\int\limits_{[0,T]^2} \psi(u)\xi(du) = \lim_{n\to\infty} \sum_{j,k=0}^{n-1} \psi(t_j,s_k)\left[\xi(t_{j+1},s_{k+1}) - \xi(t_{j+1},s_k) - \xi(t_j,s_{k+1}) + \xi(t_j,s_k)\right]$$
$$= \lim_{n\to\infty} \sum_{j,k=0}^{n-1} \psi(t_j,s_k)\left[\xi(t_{j+1},s_{k+1}) - \xi(t_{j+1},s_k) - \xi(t_j,s_{k+1}) + \xi(t_j,s_k)\right]$$
$$= \lim_{n\to\infty} \left\{ \sum_{j,k=0}^{n-1} \xi(t_j,s_k)\left[\psi(t_{j+1},s_{k+1}) - \psi(t_{j+1},s_k) - \psi(t_j,s_{k+1}) + \psi(t_j,s_k)\right] \right.$$
$$- \sum_{j=0}^{n-1} \xi(t_j,T)\left[\psi(t_j,T) - \psi(t_{j-1},T)\right]$$
$$\left. - \sum_{k=0}^{n-1} \xi(T,s_k)[\psi(T,s_k) - \psi(T,s_{k-1})] + \xi(T,T)\psi\xi(T,T) \right\},$$

where the limit is taken in the mean square sense, and $0 = t_0 < t_1 < \ldots < t_n = T$, $0 = s_0 < s_1 < \ldots < s_n = T$.

Theorem 1.16 **[24]** *Let* η *be a square integrable functional on* $\left(W(z), \Im_z, z\in[0,T]^2\right)$. *Then there exist such* $\Im_z$ *-adapted random fields* φ *and* ϕ *with*

$$\int\limits_{[0,T]^2} \varphi^2(z)dz < \infty,$$

$$\int\limits_{[0,T]^2\times[0,T]^2} \phi^2\left(z,z'\right)dzdz' < \infty,$$

that the random variable η *can be represented as*

$$\eta(z) = E\eta + \int\limits_{[0,T]^2} \varphi(z)W(ds) + \int\limits_{[0,T]^2\times[0,T]^2} \phi(z,z,')W(dz)W(dz').$$

We will need in future the following generalizations of the Ito formula [42, 47].

Theorem 1.17 *Suppose that the random field* $(\xi(z), \Im_z)$ *possesses the representation*

$$\xi(z, z'] = \int_{[z,z']} \varphi(u)du + \int_{[z,z']} \sigma(u)W(du), 0 \le z \le z', z, z' \in [0, T]^2,$$

where $(W(z), \Im_z)$ *is the Wiener field, and the random fields* $(\varphi(z), \Im_z)$ *and* $(\sigma(z), \Im_z)$ *are such that*

$$\int_{[0,T]^2} |\phi(z)|dz < \infty \text{ and } \int_{[0,T]^2} \sigma^2(z)dz < \infty$$

with probability 1. Then for any twice continuously differentiable function f(z) *we have*

$$\begin{aligned} f(\xi(t_2, s_2) - \xi(t_2, s_1)) = & f(\xi(t_1, s_2) - \xi(t_1, s_1)) \\ & + \int_{[z_1, z_2]} f'(\xi(u, s_2) - \xi(u, s_1))\sigma(u, v)W(du, dv) \\ & + \frac{1}{2} \int_{[z_1, z_2]} (f''(\xi(u, s_2) - \xi(u, s_1))\sigma^2(u, v)) \\ & + f'(\xi(u, s_2) - \xi(u, s_1))\varphi(u, v)(du, dv). \end{aligned}$$

P -almost surely.

The proof can be conducted in the same way as in the one-dimensional case, and could be found, for example, in [42].

Using the Ito formula (1.4), we can prove the following properties for two-dimensional stochastic integrals [46].

Theorem 1.18 *If* $\int_{[0,T]^2} Ef^{2m}(z)dz < \infty$ *holds true for a* $\Im_z$*-measurable function* f(z), *then for any* $0 \le \alpha, \beta \le T$ *we have*

$$E\left(\int_{[0,\alpha]\times[0,\beta]} f(z)W(dz)\right)^{2m} \le m(2m-1)^m(\alpha\beta)^{m-1} \int_{[0,\alpha]\times[0,\beta]} Ef^{2m}(z)dz.$$

Proof Applying the formula (1.4) with

$$\xi(z) = \int_{[0,z]} f(z')W(dz')$$

to the function $\varphi(x) = x^{2m}$, we obtain

$$\left[\int_{[0,z]} f\left(z'\right)W\left(dz'\right)\right]^{2m} = 2m\int_{[0,z]}\left[\int_0^u\int_0^s f(\lambda,\mu)W(d\lambda,d\mu)\right]^{2m-1} f(u,v)W(du,dv)$$
$$+\,2m\int_{[0,z]}\left[\int_0^u\int_0^s f(\lambda,\mu)W(d\lambda,d\mu)\right]^{2m-2} f^2(u,v)W(du,dv). \tag{1.5}$$

Assume first that the function $f(z)$ is piecewise constant, bounded from above by a nonrandom constant. Since for the step functions

$$f(z') = \begin{cases} f_{ik}, & z'\in[t_{i-1},t_i]\times[s_{k-1},s_k], \\ 0, & z'\notin[t_{i-1},t_i]\times[s_{k-1},s_k], \end{cases} \qquad i,k=\overline{1,n+1}.$$

the integral with respect to the Wiener measure exists, we obtain

$$E\left[\int_{[0,z]} f(z')W(dz')\right]^{4m-2} = E\left[\sum_{i,k} f_{ik}W\left(z_{i-1k-1},z_{ik}\right]\right]^{4m-2}$$
$$\le cE\sum_{i,k} f_{ik}W\left(z_{i-1k-1},z_{ik}\right].$$

Hence,

$$\int_{[0,T]^2} E\left[\int_{[0,z]} f\left(z'\right)W\left(dz'\right)\right]^{4m-2} f^2(z)dz < \infty.$$

Consequently,

$$E\left[\int_{[0,z]} f\left(z'\right)W\left(dz'\right)\right]^{2m} = 2m\int_{[0,z]}\left[\int_0^u\int_0^s f(\lambda,\mu)W(d\lambda,d\mu)\right]^{2m-1} f^2(u,v)dudv,$$

which implies that $E\left[\int_{[0,z]} f\left(z'\right)W\left(dz'\right)\right]^{2m}$ is increasing in t.

Applying the Hoelder inequality we obtain

$$E\left[\int_0^\alpha\int_0^\beta f(z')W(dz')\right]^{2m} \le m(2m-1)\left\{\int_0^\alpha\int_0^\beta E\left[\int_0^u\int_0^\beta f(\lambda,\mu)W(d\lambda,d\mu)\right]^{2m-1} dudv\right\}^{1-\frac{1}{m}} \times\left\{\int_0^\alpha\int_0^\beta Ef^{2m}(u,v)dudv\right\}^{\frac{2}{2m}}$$

and

$$E\left[\int_0^\alpha\int_0^\beta f\left(z'\right)W\left(dz'\right)\right]^{2m} \le m(2m-1)\left\{\int_0^\alpha\int_0^\beta E\left[\int_0^\alpha\int_0^\beta f(\lambda,\mu)W(d\lambda,d\mu)\right]^{2m-1} dudv\right\}^{1-\frac{1}{m}} \times\left\{\int_0^\alpha\int_0^\beta Ef^{2m}(u,v)dudv\right\}^{\frac{2}{2m}}.$$

Raising both sides to the power m and reducing by $\left(E\left[\int_0^\alpha\int_0^\beta f\left(z'\right)W\left(dz'\right)\right]^{2m}\right)^{m-1}$, we obtain the statement of the theorem for piecewise constant functions, bounded from above by a nonrandom constant. The general case is obtained by using the limit procedure. □

Theorem 1.19 [42] *Let the conditions of Theorem 1.12 be satisfied, and assume that the functions* f *and* g *are twice continuously differentiable. Then with probability 1*

(a)

$$\int_{[0,z]} f(u)W(du)\int_{[0,z]} g(u)W(du) = \int_{[0,z]} f(u)g(u)du + \int_{[0,z]} f(x,y)\left(\int_0^x\int_0^t g(\alpha,\beta)W(d\alpha,d\beta)\right)W(dx,dy) + \int_{[0,z]} g(x,y)\left(\int_0^x\int_0^t f(\alpha,\beta)W(d\alpha,d\beta)\right)W(dx,dy);$$

(b)

$$\int_{[0,z]} f(u)W(du)\int_{[0,z]} g(u)W(du) = \int_{[0,z]} f(x,y)\left(\int_0^x\int_0^t g(\alpha,\beta)d\alpha d\beta\right)W(dx,dy)$$
$$+\int_{[0,z]} g(x,y)\left(\int_0^x\int_0^t f(\alpha,\beta)W(d\alpha,d\beta)\right)dx,dy.$$

Proof To prove the first relation we apply (1.5). We obtain

$$\int_{[0,z]} f\left(z'\right)W\left(dz'\right)\int_{[0,z]} g\left(z'\right)W\left(dz'\right)$$
$$=\frac{1}{2}\left(\int_{[0,z]}\left(f\left(z'\right)+g\left(z'\right)\right)W\left(dz'\right)\right)^2$$
$$-\frac{1}{2}\left(\int_{[0,z]} f\left(z'\right)W\left(dz'\right)\right)^2-\frac{1}{2}\left(\int_{[0,z]} g\left(z'\right)W\left(dz'\right)\right)^2$$
$$=\int_0^t\int_0^s\left(\int_0^u\int_0^s (f(\lambda,\mu)+g(\lambda,\mu))W(d\lambda,d\mu)\right)(f(u,v)+g(u,v))W(du,dv)$$
$$+\frac{1}{2}\int_{[0,z]}\left(f\left(z'\right)+g\left(z'\right)\right)^2dz'-\int_0^t\int_0^s\left(\int_0^u\int_0^s f(\lambda,\mu)W(d\lambda,d\mu)\right)f(u,v)W(du,dv)$$
$$-\frac{1}{2}\int_{[0,z]} f^2\left(z'\right)dz'-\int_0^t\int_0^s\left(\int_0^u\int_0^s g(\lambda,\mu)W(d\lambda,d\mu)\right)g(u,v)W(du,dv)-\frac{1}{2}\int_{[0,z]} g^2\left(z'\right)dz'$$
$$=\int_{[0,z]} f(u)g(u)du+\int_{[0,z]} f(x,y)\left(\int_0^x\int_0^t g(\alpha,\beta)W(d\alpha,d\beta)\right)W(dx,dy)$$
$$+\int_{[0,z]} g(x,y)\left(\int_0^x\int_0^t f(\alpha,\beta)W(d\alpha,d\beta)\right)W(dx,dy).$$

Thus, (a) is proved. The property (b) can be proved similarly, using (1.5) for the right-hand side part. □

Corollary 1.3 *Under the conditions of Theorem 1.17 we have*

$$E\int_{[0,z]} f(u)W(du)\int_{[0,z]} g(u)W(du) = \int_{[0,z]} Ef(u)g(u)du.$$

Below we give a more general version of the Ito formula.

Theorem 1.20 [30, 31] *Let the conditions of Theorem 1.14 be satisfied for the random field* $(\xi(z), \Im_z)$. *Then*

$$f(\xi(z)) = f(0) + \int_{[0,z]} f'(\xi(u))\phi(u)du + \int_{[0,z]} f'(\xi(u))\sigma(u)W(du)$$
$$+ \int_{[0,z]} f''(\xi(\alpha,\beta)) \int_0^\alpha \phi(x,\beta)dx \int_0^\beta \phi(\alpha,y)dyd\alpha d\beta$$
$$+ \int_{[0,z]} f''(\xi(\alpha,\beta)) \int_0^\alpha \phi(x,\beta)dx \int_0^\beta \sigma(\alpha,y)W(d\alpha,dy)d\beta$$
$$+ \int_{[0,z]} f''(\xi(\alpha,\beta)) \int_0^\alpha \sigma(x,\beta)W(dx,d\beta) \int_0^\beta \phi(\alpha,y)dyd\alpha$$
$$+ \int_{[0,z]} f''(\xi(\alpha,\beta)) \int_0^\alpha \sigma(x,\beta)W(dx,d\beta) \int_0^\beta \sigma(\alpha,y)W(d\alpha,dy)$$
$$+\frac{1}{2} \int_{[0,z]} f'''(\xi(\alpha,\beta)) \int_0^\alpha \phi(x,\beta)dx \int_0^\beta \sigma^2(\alpha,y)dyd\alpha d\beta$$
$$+\frac{1}{2} \int_{[0,z]} f'''(\xi(\alpha,\beta)) \int_0^\alpha \sigma(x,\beta)W(dx,d\beta) \int_0^\beta \sigma^2(\alpha,y)dyd\alpha$$
$$+\frac{1}{2} \int_{[0,z]} f'''(\xi(\alpha,\beta)) \int_0^\alpha \sigma^2(x,\beta)dx \int_0^\beta \sigma(\alpha,y)W(d\alpha,dy)d\beta$$
$$+\frac{1}{4} \int_{[0,z]} f''''(\xi(\alpha,\beta)) \int_0^\alpha \sigma^2(x,\beta)dx \int_0^\beta \sigma^2(\alpha,y)dyd\alpha d\beta$$

holds true P *-almost surely for any bounded four times differentiable function* f(z).

As an example of the application of the Ito formula, we consider below the generalization of the Lévy-Doob theorem.

Theorem 1.21 [25, 42, 64] *If* μ(z) *is a square integrable strong martingale with* Eμ = 0, $E(\mu(z,z']/\Im_z^*) = (t'-t)(s'-s)$, *then* μ(z) *is a Wiener martingale.*

Proof We apply the Ito formula to the function $\exp\{i\alpha\mu_{(z_0,z]}\}$. Denote by *j(α,z)* the characteristic function of the conditional distribution $\mu(z_0, z]$, $z \geq z_0$,

$$j(\alpha,z) = E\left(\exp\{i\alpha\mu_{(z_0,z]}\}/\Im_{z_0}^*\right).$$

We obtain

$$j(\alpha,z) = 1 - \int\limits_{(z_0,z]} \left(\frac{\alpha^2}{2} j(\alpha,z') + \frac{\alpha^4}{4} j(\alpha,z')(t'-t_0)(s'-s_0)\right) dt' ds'.$$

Let $\mu_1(z) := \mu(z_0, z]$, $z \geq z_0$. Therefore,

$$E\left(\int\limits_{(z_0,z]} j(\alpha,z')\mu(dt',s')\mu(t',ds')/\Im^*_{z_0}\right) = \int\limits_{(z_0,z]} j(\alpha,z')\mu_1(dt',s')\mu 1(t',ds').$$

This equation is equivalent to the following one:

$$\frac{\partial^2 j}{\partial t \partial s} = -\left(\frac{\alpha^2}{2} - \frac{\alpha^4}{4}(t-t_0)(s-s_0)\right) j, \quad j(t_0,s) = j(t,s_0) = 1,$$

whose solution is $j(\alpha,z) = e^{-\frac{\alpha^2}{2}(t-t_0)(s-s_0)}$. Thus, the random variable $\mu(z_0, z]$ does not depend on the σ-algebra $\Im^*_{z_0}$ and has the Gaussian distribution with mean 0 and variance $(t - t_0)(s - s_0)$. □

1.3 Stochastic Measures and Integrals for Nonrandom Functions

In applications, a lot of models use the integrals of the form $\int f(z)\xi(dz)$, where $f(z)$ is a nonrandom function, and $\xi(z)$ is a random field. Since in general $\xi(z)$ is not expected to be of bounded variation, such an integral cannot be treated in the Stieltjes or Lebesgue-Stieltjes sense. Nevertheless, it can be determined in such a way that it inherits the properties of the Lebesgue-Stieltjes integral.

Following [28], we consider the probability space $(\Omega, \Im, P)$, $L_2 = L_2(\Omega, \Im, P)$, S is some set, $\mathfrak{R}$ is the semiring generated by sub-sets of S. For any $\Delta \in \mathfrak{R}$ consider the complex-valued variable $\mu(\Delta)$ satisfying the following conditions:

(1) $\mu(\Delta) \in L_2(\Omega, \Im, P)$, $\mu(\varnothing) = 0$,
(2) $\mu(\Delta_1 \cup \Delta_2) = \mu(\Delta_1) + \mu(\Delta_1)$ (mod P), if $\Delta_1 \cap \Delta_2 = \varnothing$,
(3) $E\mu(\Delta_1)\overline{\mu(\Delta_2)} = m(\Delta_1 \cap \Delta_2)$,
where $m(\Delta)$ is some set function on $\mathfrak{R}$.

We call the family of random variables $\{\mu(\Delta)\}$, $\Delta \in \mathfrak{R}$, the elementary orthogonal stochastic measure, and respective function $m(\Delta)$ its structure function. In this context orthogonality means that

(4) if $\Delta_1 \cap \Delta_2 = \varnothing$, then the variables $\mu(\Delta_1)$ and $\mu(\Delta_2)$ are orthogonal.

By definition, the function $m(\Delta)$ is nonnegative: $m(\Delta) = E|\mu(\Delta)|^2 \geq 0$, $m(\varnothing) = 0$, and additive: if $\Delta_1 \cap \Delta_2 = \varnothing$, then

$$\begin{aligned} m(\Delta_1 \cup \Delta_2) &= E|\mu(\Delta_1) + \mu(\Delta_1)|^2 = m(\Delta_1) + m(\Delta_2) + 2m(\Delta_1 \cap \Delta_2) \\ &= m(\Delta_1) + m(\Delta_2). \end{aligned}$$

Denote by $L_0\{\mathfrak{R}\}$ the class of all simple functions $f(x)$, $f(x) = \sum_{k=1}^{n} c_k \chi_{\Delta_k}(x)$, $\Delta_k \in \mathfrak{R}$, $k = 1, \ldots, n$, where n is any natural number, and $\chi_A(x)$ is the indicator of a set A. For functions $f \in L_0\{\mathfrak{R}\}$ we define the stochastic integral with respect to the elementary stochastic measure μ by the formula

$$I := \int f(x)\mu(dx) = \sum_{k=1}^{n} c_k \mu(\Delta_k). \tag{1.6}$$

Since $\mathfrak{R}$ is a semiring, any function from $L_0\{\mathfrak{R}\}$ can be represented as linear combination of indicators of sets from $\mathfrak{R}$. Let $f, g \in L_0\{\mathfrak{R}\}$, and put

$$g(x) = \sum_{k=1}^{n} d_k \chi_{\Delta_k}(x),$$

where $\Delta_k \cap \Delta_l = \varnothing$ if $k \neq l$.

It follows from the orthogonality of the measure that

$$E\left(\int f(x)\mu(dx) \overline{\int g(x)\mu(dx)}\right) = \sum_{k=1}^{n} c_k \overline{d}_k m(\Delta_k).$$

Suppose that the elementary measure m is sub-additive, and hence can be extended to a full measure on $(E, \aleph, m)$. Then $L_0\{\mathfrak{R}\}$ is a linear subspace of the Hilbert space $L_2\{m\} = L_2(E, \aleph, m)$, and $L_2\{\mathfrak{R}\}$ is the closure of $L_0\{\mathfrak{R}\}$ in the topology generated by the scalar product $(f, g) = \int f(x)\overline{g(x)}\, m(dx)$. At the same time,

$$E\left(\int f(x)\mu(dx) \overline{\int g(x)\mu(dx)}\right) = \int f(x)\overline{g(x)}\, m(dx) \tag{1.7}$$

for any pair of functions $f, g \in L_0\{\mathfrak{R}\}$.

Denote by $\widetilde{L}_0\{\mu\}$ the linear span of the family of random variables $\{\mu(\Delta)\}, \Delta \in \mathfrak{R}$ (that is, $\widetilde{L}_0\{\mu\}$ is the set of random variables represented in the form (1.6)), and denote by $\widetilde{L}_2\{\mu\}$ the closure of $\widetilde{L}_0\{\mu\}$ in Hilbert space $L_2(\Omega, \mathfrak{I}, P)$. Relation (1.6) sets the isometric correspondence $I = \psi(f)$ between $L_0\{\mathfrak{R}\}$ and $\widetilde{L}_0\{\mu\}$, which can be extended [63] to the isometric correspondence between $L_2\{\mathfrak{R}\}$ and $\widetilde{L}_2\{\mu\}$. If $I = \psi(f), f \in L_2\{\mathfrak{R}\}$, we define

$$I = \psi(f) = \int f(x)\mu(dx) \tag{1.8}$$

and call the random variable I the stochastic integral of the function $f(x)$ with respect to the measure μ. The properties of such an integral are given in the theorem below.

Theorem 1.22 [25]

(a) for a simple function the stochastic integral can be defined by (1.6);
(b) for any $f, g \in L_2(E,\aleph,m)$ equality (1.7) holds true;
(c) $\int[\alpha f(x) + \beta g(x)]\mu(dx) = \alpha \int f(x)\mu(dx) + \beta \int g(x)\mu(dx)$;
(d) for any sequence of functions $f_n \in L_2(E,\aleph,m)$ such that

$$\int |f(x) - f_n(x)|^2 m(dx) \to 0, \quad n \to \infty, \tag{1.9}$$

we have $\int f(x)\mu(dx) = \lim \int f_n(x)\mu(dx)$.

Remark 1.13 In particular, if $f_n(x)$, $n \geq 1$, is a sequence of simple functions, $f_n(x) = \sum_{k=1}^{m_n} c_k^{(n)} \chi_{\Delta_k^{(n)}}(x)$, $\Delta_k^{(n)} \in \mathfrak{R}$, $n = 1, 2, \ldots$, and (1.9) is satisfied, then

$$\int f(x)\mu(dx) = \lim \sum_{k=1}^{m_n} c_k^{(n)} \mu\left(\Delta_k^{(n)}\right).$$

The existence of the sequence of simple functions which approximates an arbitrary $L_2(E,\aleph,m)$-function, was proved, for instance, in [35]. Thus, the stochastic integral can be considered as the mean square limit of appropriate integral sums.

Denote by $\aleph_0$ the class of all sets $A \in \aleph$, for which $m(\Delta) < \infty$. Define a random function $\widetilde{\mu}(A)$ of the set A:

$$\widetilde{\mu}(A) = \int \chi_A(x)\mu(dx) = \int_A \mu(dx). \tag{1.10}$$

This function possesses the following properties:

(a) $\widetilde{\mu}(A)$ is well defined on sets from $\aleph_0$

(b) if $A_n \in \aleph_0$, $n = 1, 2, \ldots$, $A_0 = \bigcup_{n=1}^{\infty} A_n$, $A_k \cap A_l = \varnothing$, $k \neq l$, then $\widetilde{\mu}(A_0) = \sum_{n=1}^{\infty} \widetilde{\mu}(A_n) \pmod P$;

(c) $E\widetilde{\mu}(A)\widetilde{\mu}(B) = m(A \cap B)$, $A, B \in \aleph_0$;
(d) $\widetilde{\mu}(\Delta) = \mu(\Delta)$ as $\Delta \in \mathfrak{R}$.

A random set function which satisfies the conditions (a), (b), (c) is called the stochastic orthogonal measure. Property (d) means that $\widetilde{\mu}(A)$ is the extension of the elementary stochastic measure μ.

Corollary 1.4 *If the structure function of an elementary stochastic measure* μ *is sub-additive, then* μ *can be extended to a stochastic measure.*

Remark 1.14 Since $\widetilde{L}_2\{\mu\} = \widetilde{L}_2\{\widetilde{\mu}\}$,

$$\int f(x)\mu(dx) = \int f(x)\widetilde{\mu}(dx).$$

In the future, we identify a stochastic integral with respect to the elementary orthogonal measure μ with sub-additive structure function and the stochastic integral with respect to the measure $\widetilde{\mu}$, defined in (1.10).

Let μ be the orthogonal stochastic measure with structure function m, which is the full measure on $\{E, \aleph\}$ and $g(z) \in \widetilde{L}_2\{m\}$. Put

$$\lambda(A) = \int \chi_A(x) g(x)\mu(dx), A \in \aleph$$

Then

$$E\lambda(A)\overline{\lambda(B)} = \int \chi_A(x)\chi_B(x) g^2(x) m(dx) = \int\limits_{A\cap B} |g(x)|^2 m(dx).$$

If we introduce a new measure

$$l(A) = \int\limits_A |g(x)|^2 m(dx)$$

on $\aleph$, then $\lambda(A)$ will be the orthogonal stochastic measure with the structure function $l(A)$, $A \in \aleph$.

Lemma 1.3 *If* $f \in \widetilde{L}_2\{l\}$, *then* $fg \in \widetilde{L}_2\{m\}$ *and* $\int f(x)\lambda(dx) = \int f(x)g(x)\mu(dx)$.

Proof The statement of the lemma is obvious for simple functions $f(x) = \sum_k c_k \chi_{A_k}(x)$, $A_k \in \aleph$. If $f_k(x)$ is a fundamental sequence of simple functions in $\widetilde{L}_2\{l\}$, then

$$\left\| \int f_n(x)\lambda(dx) - \int f_{n+m}(x)\lambda(dx) \right\|^2 = \int \left| f_n(x) - f_{n+m}(x) \right|^2 l(dx)$$
$$= \int \left| f_n(x) - f_{n+m}(x) \right|^2 |g(x)|^2 m(dx).$$

That is, the sequence $f_n(x)g(x)$ is fundamental in $\widetilde{L}_2\{m\}$. Passing to the limit as $n \to \infty$ in

$$\int f_n(x)\lambda(dx) = \int f_n(x)g(x)\mu(dx),$$

we obtain the statement of the lemma. □

Lemma 1.4 *If* $\lambda(A) = \int \chi_A(x)g(x)\mu(dx)$ *and* $g \in \widetilde{L}_2\{m\}$, *then for any* $A \in \aleph_0$

$$\mu(A) = \int \frac{1}{g(x)} \chi_A(x)\lambda(dx).$$

Proof First, $g(x) = 0$ on the set of l -measure 0, implying $\dfrac{1}{g(x)} \neq \infty \pmod{l}$. Moreover, if $A \in \aleph_0$, then

$$\int \frac{1}{|g(x)|^2} \chi_A(x) l(dx) = \int_A \frac{1}{|g(x)|^2} |g(x)|^2 m(dx) = m(A) < \infty.$$

From Lemma 1.3 we derive

$$\int \frac{1}{g(x)} \chi_A(x) l(dx) = \int \frac{1}{g(x)} \chi_A(x) g(x)\mu(dx) = \mu(A).$$ □

Let D be a rectangle (finite or infinite) on the plane, $\mathcal{B}$ is the σ-algebra of Lebesgue measurable subsets of D, and l is the Lebesgue measure. Suppose that the function $g(z,x)$ is $\mathcal{B} \times \aleph$ -measurable, $g(z,x) \in \widetilde{L}_2\{l \times m\}$ and $g(z,x) \in \widetilde{L}_2\{m\}$ for arbitrary $z \in D$. Consider the stochastic integral

$$\xi(z) = \int g(z,x)\mu(dx) \tag{1.11}$$

which is well defined for any z with probability 1.

Lemma 1.5 *The stochastic integral (1.11) can be determined as a function of z in such a way that the random field* $\xi(z)$ *is measurable.*

Proof If $g(z,x) = \sum c_k \chi_{B_k}(z)\chi_{A_k}(x)$, $B_k \in \mathcal{B}$, $A_k \in \aleph$, then $\xi(z) = \sum c_k \chi_{B_k}(z)$ $\mu(A_k)$ is $\mathcal{B} \times \Im$-measurable function of variables $z \in D$ and $\omega \in \Im$. In the general case we can build a sequence of simple functions $g_n(z,x)$ such that

$$\int\int |g(z,x) - g_n(z,x)|^2 m(dx)dz \to 0, \text{ as } n \to \infty.$$

Let $\xi_n(z)$ be a sequence of random fields constructed as in (1.11) with $g = g_n$. Then there exists a field $\widetilde{\xi}(z)$ such that $\int E\left|\widetilde{\xi}(z) - \xi_n(z)\right|^2 dz \to 0$ as $n \to \infty$, and $\widetilde{\xi}(z)$ is a $\mathcal{B}_k \in \mathcal{B} \times \Im$-measurable function.

On the other hand,

$$\int E|\xi(z) - \xi_n(z)|^2 dz = \iint |g(z,x) - g_n(z,x)|^2 m(dx)dz \to 0,$$

implying that the random fields $\xi(z)$ and $\widetilde{\xi}(z)$ coincide for almost all z. Put

$$\hat{\xi}(z) = \begin{cases} \widetilde{\xi}(z), & P\left\{\xi(z) \neq \widetilde{\xi}(z)\right\} = 0, \\ \xi(z), & P\left\{\xi(z) \neq \widetilde{\xi}(z)\right\} > 0. \end{cases}$$

The field $\hat{\xi}(z)$ is measurable (since $\hat{\xi}(z)$ differs from a $B_k \in \mathcal{B} \times \Im$-measurable function $\widetilde{\xi}(z)$ only on a set of measure 0) and is stochastically equivalent to $\xi(z)$. □

Everywhere below we assume that the random field determined by a stochastic integral of the form (1.11) and satisfying the conditions of Lemma 1.3 is measurable.

Lemma 1.6 *If* g(z,x) *and* h(z) *are Borel functions,* I = [a_1,b_1] × [a_2,b_2] ⊂ D,

$$\int_I \int_{-\infty}^{\infty} |g(z,x)|^2 dz m(dx) < \infty, \int_I |h(z)|^2 dz < \infty,$$

and μ *is an orthogonal stochastic measure on* $\mathcal{B}$, *then*

$$\int_I h(z) \int_{-\infty}^{\infty} g(z,x)\mu(dx)dz = \int_{-\infty}^{\infty} g_1(x)\mu(dx), \tag{1.12}$$

where $g_1(x) = \int_I h(z)g(z,x)dz.$

Proof Let us estimate the square of the mean in the left-hand side of (1.12). We have

$$\int_I \int_I h(z_1)\overline{h(z_2)} \int_{-\infty}^{\infty} g(z_1, x)\overline{g(z_2, x)}\, m(dx)dz_1 dz_2$$

$$= \int_{-\infty}^{\infty} \left| \int_I h(z)g(z, x)dz \right|^2 m(dx)$$

$$\leq \int_I |h(z)|^2 dz \int_{-\infty}^{\infty} \int_I |g(z, x)|^2 dzm(dx).$$

We obtain similar inequality for the right-hand side of relation (1.12) as well. Thus, we can take in (1.12) a sequence $g_n(z,x)$ converging in $\widetilde{L}_2\{l \otimes m\}$, and pass to the limit. Moreover, the set of functions $g(z,x)$, for which (1.12) holds true, is linear and contains all functions of the form $g(z, x) = \sum c_k \chi_{B_k}(z)\chi_{A_k}(x)$, Hence, it contains all functions of $L_2\{l \otimes m\}$. □

Lemma 1.7 *If the conditions of Lemma 1.3 hold true for each bounded* I ⊂ D, I = *[*a_1,b_1*]* × *[*a_2,b_2*], and the integral*

$$\int_{R^2} h(z)g(z, x)dz = \lim_{\substack{a_i \to -\infty \\ b_i \to \infty}} \int_I h(z)g(z, x)dz, \quad i = 1, 2,$$

exists in the sense of convergence in L_2*{*m*}, then*

$$\int_{R^2} h(z) \int_{-\infty}^{\infty} g(z, x)\mu(dx)dz = \int_{-\infty}^{\infty} g_1(x)\mu(dx), \quad g_1(x) = \int_I h(z)g(z, x)dz.$$

The proof follows immediately from the fact that $\int_{R^2} h(z) \int_{-\infty}^{\infty} g(z, x)\mu(dx)dz$ *is the mean square limit of (1.12), and one can pass to the limit under the sign of the stochastic integral in the right-hand side of (1.12).*

Chapter 2
Stochastic Differential Equations on the Plane

In this chapter we investigate diffusion-type fields and Ito fields on the plane, two-parameter version of the Girsanov theorem, weak and strong solutions of stochastic differential equations on the plane, and the probability measures generated by stochastic fields. The results presented in this chapter are published in [10, 12, 14, 16, 25, 42, 44, 45, 47, 48, 65, 71].

2.1 Ito and Diffusion-Type Stochastic Fields

Denote by $\left(C[0,T]^2,\mathcal{B}\right)$ the space of continuous functions on $[0,T]^2$. We denote by $\|\cdot\|$ the uniform norm in $C[0,T]^2$ and by $\|\cdot\|_z$ —the uniform norm in $C[0,z]^2$. In this chapter we suppose that the Cairoli-Walsh condition F4 is fulfilled. Let $f\in C[0,T]^2$ and assume in addition that satisfies the boundary conditions $f(0,s)=f(t,0)=0$. In the space $\left(C[0,T]^2,\mathcal{B}\right)$ we also define the σ-algebras $\mathcal{B}=\sigma\left\{f(z),z\in[0,T]^2\right\}$ and $\mathcal{B}_z=\sigma\{f(z'),z'\leq z\}$.

Definition 2.1 We say that a measurable with respect to both variables functional $\varphi(z,f)$, $z\in[0,T]^2$, $f\in C[0,T]^2$, does not depend on the future, if for any z the functional $\varphi(z,f)$ is $\mathcal{B}_z$-measurable.

Definition 2.2 A continuous stochastic field $(\xi(z),\mathfrak{I}_z)$ defined on the probability space $(\Omega,\mathfrak{I},P)$ is called

(a) An Ito field (with respect to the Wiener field $(W(z),\mathfrak{I}_z)$), if there exist random fields $(A(z),\mathfrak{I}_z)$ and $(B(z),\mathfrak{I}_z)$, $z\in[0,T]^2$, such that

$$\int_{[0,T]^2}|A(z)|dz<\infty\quad\text{and}\quad\int_{[0,T]^2}B^2(z)dz<\infty\quad P\text{-a.s.},\tag{2.1}$$

P.S. Knopov and O.N. Deriyeva, *Estimation and Control Problems for Stochastic Partial Differential Equations*, Springer Optimization and Its Applications 83,
DOI 10.1007/978-1-4614-8286-4_2, © Springer Science+Business Media New York 2013

and for any $z \in [0,T]^2$

$$\xi(z) = \int_{[0,z]} A(u)du + \int_{[0,z]} B(u)W(du)$$

with probability 1.

(b) A diffusion-type field (with respect to the Wiener field $(W(z), \Im_z)$), if there exist functionals $a(z,f)$ and $b(z,f)$, $z \in [0,T]^2$, $f \in C[0,T]^2$, which do not depend on the future,

$$\int_{[0,T]^2} |a(z,\xi)|dz < \infty \quad \text{and} \quad \int_{[0,T]^2} b^2(z,\xi)dz < \infty \quad P\text{-a.s.},$$

and for any $z \in [0,T]^2$

$$\xi(z) = \int_{[0,z]} a(u,\xi)du + \int_{[0,z]} b(u,\xi)W(du)$$

with probability 1.

Let $\overline{f}^{-1}(z) := \begin{cases} f^{-1}(z), & f(z) \neq 0, \\ 0, & f(z) = 0. \end{cases}$

Often Ito fields can be represented as diffusion-type fields, probably with respect to another Wiener field.

Theorem 2.1 [14, 16] *Assume that $(\xi(z), \Im_z)$ is an Ito field and that stochastic fields $(A(z), \Im_z)$ and $(B(z), \Im_z)$, $z \in [0,T]^2$, satisfy condition (2.1) as well as the conditions below:*

(a) For almost all $z \in [0,T]^2$ the inequality $B^2(z) > 0$ holds true with probability 1.

(b) $\int_{[0,T]^2} E|A(z)|dz < \infty$.

(c) $\int_{[0,T]^2} E\left|\overline{B}^{-1}(z)\right||A(z)|dz < \infty$.

Then in the space $(C[0,T]^2, \mathcal{B})$ *there exist measurable functionals $\alpha(z,f)$ and $\beta(z,f)$,*

$$\alpha(z,\xi) = E(A(z)/\Im_z^\xi), \beta(z,\xi) = \left(E(B^2(z)/\Im_z^\xi)\right)^{1/2},$$

$\Im_z^\xi = \sigma\left\{\xi(z'), z' \in [0,T]^2, z' \leq z\right\}$ *and a Wiener field* $(\hat{W}(z), \Im_z^\xi)$, *such that*

$$\xi(z) = \int_{[0,z]} \alpha(u,\xi)du + \int_{[0,z]} \beta(u,\xi)\hat{W}(du)$$

with probability 1.

Proof On the space $\left(C[0,T]^2, \mathcal{B}\right)$ we construct a stochastic field $\zeta(z,\omega)$, $z \in [0,T]^2$, as $\zeta(z,\omega) = E\left(A(z)/\mathfrak{I}_z^\xi\right)$. Since the field is $\zeta(z,\omega)$ obviously measurable, $\mathfrak{I}_z^\xi$-adapted, then from condition (b) we obtain for any $z \in [0,T]^2$ (see [52] for the similar trick)

$$\{(\omega, z' \le z) : \xi(z',\omega) \in \mathrm{B}\} \in \mathfrak{I}_z \times \mathcal{B}_z,$$

where B is a Borel set in R, $\mathcal{B}_z$ is a σ-algebra of Borel sets on $[0,z]$. This means that the field $\zeta(z,\omega)$, $z \in [0,T]^2$, has a progressively measurable modification. Therefore, without loss of generality, we assume that the field $\zeta(z)$ is progressively measurable. Then for any $z' \in [0,T]^2$ the stochastic field $\zeta(z \wedge z', \omega)$ (as a function of z) is measurable in the space $\left([0,T]^2 \times \Omega,\ \mathcal{B}_z \times \mathfrak{I}_z^\xi,\ \mu \times P\right)$, where μ is the Lebesgue measure on $[0,z]$. Therefore there exists a $\mathfrak{I}_z^\xi$-adapted functional $\alpha_{z_0}(z,f)$ such that

$$\mu \times P\{(z,\omega) : \zeta(\min(z,z_0),\omega) \ne \alpha_{z_0}(z,\xi(\omega))\} = 0.$$

For any $n \ge 1$ we put $z_{k,j,n} = (2^{-n}k,\ 2^{-n}j)$ and

$$I_{k,j,n} = \left(z_{k-1,j-1,n}, z_{k,j-1,n}\right] \times \left(z_{k,j-1,n}, z_{k,j,n}\right].$$

Then we build piecewise constant functions

$$\alpha^{(n)}(z,f) = \alpha_0(0,f)\chi_{\{0\}}(z) + \sum_{k=1}^{2^n}\sum_{j=1}^{2^n} \alpha_{z_{k,j,n}}\chi_{I_{k,j,n}}(z),$$

and put

$$\alpha(z,f) = \lim_{n\to\infty} \alpha^{(n)}(z,f).$$

For any $z \in [0,T]^2$ the functional $\alpha(z,f)$ is $\mathcal{B}_z$-measurable. Moreover, for any $\varepsilon > 0$

$$\{(z,\omega) : |\alpha(z,\xi(\omega)) - \zeta(z,\omega)| > \varepsilon\} \subset \left\{(z,\omega) : \left|\alpha(z,\xi(\omega)) - \alpha^{(n)}(z,\xi(\omega))\right| > \varepsilon/2\right\}$$
$$\cup \left\{(0,\omega) : \left|\alpha^0(z,\xi(\omega)) - \zeta(z,\omega)\right| > \varepsilon/2\right\}$$
$$\cup \left\{\bigcup_{k=1}^{2^n} \left\{(z,\omega) : z \in I_{k,j,n}, \left|\alpha_{z_{k,j,n}}(z,\xi(\omega)) - \zeta\left(\min(z,z_{k,j,n})\right)\right| > \varepsilon/2\right\}\right\}.$$

Therefore,

$$\mu \times P\{(z,\omega) : |\alpha(z,\xi(\omega)) - \zeta(z,\omega)| > \varepsilon\}$$
$$\leq \mu \times P\Big\{(z,\omega) : \big|\alpha(z,\xi(\omega)) - \alpha^{(n)}(z,\xi(\omega))\big| > \varepsilon/2\Big\}.$$

As $\alpha(z,f) = \overline{\lim}_{n\to\infty}\alpha^{(n)}(z,f)$, there exists a subsequence $\{n_j\}$, $j = 1, 2, \ldots$ such that

$$\lim_{n_j\to\infty} \mu \times P\Big\{(z,\omega) : \Big|\alpha(z,\xi(\omega)) - \alpha^{(n_j)}(z,\xi(\omega))\Big| > \varepsilon/2\Big\} = 0$$

Hence, for any $\varepsilon > 0$

$$\lim_{n_j\to\infty} \mu \times P\{(z,\omega) : |\alpha(z,\xi(\omega)) - \zeta(z,\omega)| > \varepsilon/2\} = 0.$$

Now we need to construct the functional $\beta(z,\xi)$. For this purpose we consider the partition of the set $(0,z]$, such that $0 = t_0^{(n)} \leq t_1^{(n)} \leq \ldots \leq t_n^{(n)} = t$, and $0 = s_0^{(n)} \leq s_1^{(n)} \leq \ldots \leq s_n^{(n)} = s$. Let $z_{ij}^{(n)} = (t_i^{(n)}, s_j^{(n)})$, $I_{i,j,n} = (t_i^{(n)}, t_{i+1}^{(n)}] \times (s_j^{(n)}, s_{j+1}^{(n)}]$ and assume that diam $I_{i,j,n} \to 0$, $n \to \infty$. The increments on these intervals satisfy

$$\sum_{i,j=0}^{n-1}\left(\xi\big(z_{ij}^{(n)}, z_{i+1,j+1}^{(n)}\big]\right)^2 = \sum_{i,j=0}^{n-1}\left(\int_{I_{i,j,n}} A(u)du\right)^2 + \sum_{i,j=0}^{n-1}\left(\int_{I_{i,j,n}} B(u)W(du)\right)^2$$
$$+ 2\sum_{i,j=0}^{n-1}\left(\int_{I_{i,j,n}} B(u)W(du)\right) \times \left(\int_{I_{i,j,n}} A(u)du\right).$$

Performing easy calculations we obtain

$$\sum_{i,j=0}^{n-1}\left(\int_{I_{i,j,n}} A(u)du\right)^2 \leq \max_{i,j}\left|\int_{I_{i,j,n}} A(u)du\right|\left|\int_{[0,T]^2} A(u)du\right| \to 0, \quad n \to \infty$$

and

$$\left|\sum_{i,j=0}^{n-1}\left(\int_{I_{i,j,n}} B(u)W(du)\right) \times \left(\int_{I_{i,j,n}} A(u)du\right)\right|$$
$$\leq \max_{i,j}\left|\int_{I_{i,j,n}} B(u)W(du)\right| \times \left|\int_{[0,T]^2} A(u)du\right| \to 0, \quad n \to \infty.$$

Taking into account Theorem 1.17 we get

$$\sum_{i,j=0}^{n-1}\left(\int_{I_{i,j,n}} B(u)W(du)\right)^2 = \sum_{i,j=0}^{n-1}\left(\int_{I_{i,j,n}} B^2(u)du\right)$$
$$+2\sum_{i,j=0}^{n-1}\left(\int_{I_{i,j,n}}\int_{\left[z_{ij}^{(n)},z'\right]} B(u)W(du)B(z')W(dz')\right)^2$$
$$=\int_{[0,z]} B^2(u)du + 2\int_{[0,z]} f_n(u)B(u)W(du),$$

where $f_n(z')$ is defined by

$$f_n(z') = \int_{\left[z_{ij}^{(n)},z,'\right]} B(u)W(du), \quad t_i^{(n)} \leq t' \leq t_{i+1}^{(n)}, \quad s_j^{(n)} \leq s' \leq s_{j+1}^{(n)},$$

and

$$\int_{[0,T]^2} f_n^2(u)B^2(u)du \leq \max_{i,j}\ \sup_{I_{ij}} f_n^2(u)\int_{[0,T]^2} B^2(u)du,$$

Therefore,
$$\sum_{i,j=0}^{n-1}\left(\xi\left(z_{ij}^{(n)}, z_{i+1,j+1}^{(n)}\right]\right)^2 \rightarrow \int_{[0,z]} B^2(u)du \text{ as } n\rightarrow\infty \text{ with probability 1.}$$
Thus, for any $z \in [0,T]^2$ the stochastic field $\int_{[0,z]}B^2(u)du$ is $\Im_z^\xi$-measurable, and repeating the calculations for deriving the expression for α, it is easy to show the existence of β.

Consider now the stochastic field $\left(\hat{W}(z), \Im_z^\xi\right)$, $z \in [0,T]^2$, given by

$$\hat{W}(z) = \int_{[0,z]} \overline{\beta}^{-1}(u,\xi)\xi(du) - \int_{[0,z]} \overline{\beta}^{-1}(u,\xi)\alpha(u,\xi)du$$

and show that it is a Wiener field. By definition of ξ we have

$$\hat{W}(z) = \int_{[0,z]} \overline{\beta}^{-1}(u,\xi)B(u)W(du) + \int_{[0,z]} \overline{\beta}^{-1}(u,\xi)(A(u)-\alpha(u,\xi))du,$$

where the existence of these integrals is guaranteed by conditions (c) and (2.1). Condition (a) guarantees that $B(z)\overline{B}^{-1}(z) = 1$ P-a.s. For any $0 \leq t \leq t' \leq T$, $0 \leq s \leq s' \leq T$, $-\infty < \theta < \infty$ we have by the Ito formula

$$\begin{aligned}
&\exp\{i\theta(\hat{W}(t,s') - \hat{W}(t,s))\}\\
&= 1 + i\theta \int_{[0,t]\times[s,s']} \exp\{i\theta(\hat{W}(t,y) - \hat{W}(t,s))\}\overline{\beta}^{-1}(x,y,\xi)B(x,y)W(dx,dy)\\
&\quad + i\theta \int_{[0,t]\times[s,s']} \exp\{i\theta(\hat{W}(t,y) - \hat{W}(t,s))\}\overline{\beta}^{-1}(x,y,\xi)[A(x,y) - \alpha(x,y,\xi)]dxdy\\
&\quad - \frac{\theta^2}{2}\int_{[0,t]\times[s,s']} \exp\{i\theta(\hat{W}(t,y) - \hat{W}(t,s))\}dxdy
\end{aligned}$$

Then, taking the conditional expectation, we get

$$E(\exp\{i\theta(\hat{W}(t,s') - \hat{W}(t,s))\}/\Im_z^{\xi 2}) = \exp\left(-\frac{\theta^2}{2}t(s'-s)\right).$$

Similarly,

$$E(\exp\{i\theta(\hat{W}(t',s) - \hat{W}(t,s))\}/\Im_z^{\xi 1}) = \exp\left(-\frac{\theta^2}{2}(t'-t)s\right).$$

Therefore, the stochastic field $(\hat{W}(z), \Im_z^{\xi})$ is a Wiener field (see Theorem 1.3).

Now we prove that $\xi(z)$ is a diffusion-type field with respect to $(\hat{W}(z), \Im_z^{\xi})$. It is easy to see that

$$\begin{aligned}
\int_{[0,z]} \beta(u,\xi)\hat{W}(du) &= \int_{[0,z]} \beta(u,\xi)\overline{\beta}^{-1}(u,\xi)\xi(du) - \int_{[0,z]} \beta(u,\xi)\overline{\beta}^{-1}(u,\xi)\alpha(u,\xi)du\\
&= \xi(z) - \xi(0) - \int_{[0,z]} \alpha(u,\xi)du + \eta(z)
\end{aligned}$$

where

$$\eta(z) = \int_{[0,z]} \left[1 - \beta(u,\xi)\overline{\beta}^{-1}(u,\xi)\right][\xi(du) - \alpha(u,\xi)du]$$

Now we show that $\eta(z)$ is a martingale.

$$\begin{aligned}
\eta(z) = &\int_{[0,z]} \left[1 - \beta(u,\xi)\overline{\beta}^{-1}(u,\xi)\right]B(u)W(du)\\
&+ \int_{[0,z]} \left[1 - \beta(u,\xi)\overline{\beta}^{-1}(u,\xi)\right][A(u) - \alpha(u,\xi)]du
\end{aligned}$$

Condition (a) guarantees that

$$\left[1 - \beta(u,\xi)\overline{\beta}^{-1}(u,\xi)\right]B(u) = 0$$

and

$$E\left(\int_{[0,z]}\left[1-\beta(u,\xi)\overline{\beta}^{-1}(u,\xi)\right]B(u)W(du)\right)^2=0.$$

Further, recall that

$$E\{(A(z')-\alpha((z'),\xi))/\Im_z^\xi\}=E\{(A(z')-E[A(z')/\Im_z^\xi])/\Im_z^\xi\}=0,$$

a.s. for arbitrary $z\le z'$, implying $E\{\eta(z')/\Im_z^\xi\}=\eta(z)$. Conditions (a)–(c) guarantee that

$$P\left(\int_{[0,T]^2}\left|\left[1-\beta((z),\xi)\overline{\beta}^{-1}((z),\xi)\right][A(z)-\alpha((z),\xi)]\right|dz<\infty\right)=1.$$

Denote $\varphi(z):=\left[1-\beta((z),\xi)\overline{\beta}^{-1}((z),\xi)\right][A(z)-\alpha((z),\xi)]$ and put

$$\tau_N:=\inf\left\{t\le t':\int_0^t\int_0^{s'}|\varphi(z)|dz\ge N\right\}\quad\text{and}\quad\tau_N:=t'\quad\text{if}\quad\int_0^{t'}\int_0^{s'}|\varphi(z)|dz<N.$$

Further, let $\chi^{(N)}(t):=\chi_{\{t\le\tau_N\}}$ and $\varphi^{(N)}(t):=\int_0^t\int_0^{s'}\chi^{(N)}(x)\varphi(x,y)dxdy$. The one-parameter stochastic process $\varphi^{(N)}(t),\ t\le t'$ with filtration $\Im_t^{(N)}=\Im_{(t\wedge\tau_N,s')}$ is the square integrable martingale, hence (see [54])

$$E\left(\varphi^{(N)}(t)\right)^2=\lim_{n\to\infty}\sum_{i=0}^{n-1}E\left(\varphi^{(N)}(t_{i+1})-\varphi_i^{(N)}(t)\right)^2$$

where $0=t_0<\ldots<t_n\le t'$ and $\max_i|t_{i+1}-t_i|\to 0$ as $n\to\infty$. We have

$$\begin{aligned}
E\left(\varphi^{(N)}(t)\right)^2&=\lim_{n\to\infty}\sum_{i=0}^{n-1}E\left(\int_{t_i}^{t_{i+1}}\int_0^{s'}\chi^{(N)}(x)\varphi(x,y)dxdy\right)^2\\
&\le\underline{\lim}_{n\to\infty}E\left\{\max_{i\le n-1}\int_{t_i}^{t_{i+1}}\int_0^{s'}\chi^{(N)}(x)|\varphi(x,y)|dxdy\sum_{i=0}^{n-1}E\int_{t_i}^{t_{i+1}}\int_0^{s'}\chi^{(N)}(x)|\varphi(x,y)|dxdy\right\}\\
&\le\underline{\lim}_{n\to\infty}E\left\{\max_{i\le n-1}\int_{t_i}^{t_{i+1}}\int_0^{s'}\chi^{(N)}(x)|\varphi(x,y)|dxdy\int_0^{t'}\int_0^{s'}\chi^{(N)}(x)|\varphi(x,y)|dxdy\right\}\\
&\le N\,\underline{\lim}_{n\to\infty}E\left\{\max_{i\le n-1}\int_{t_i}^{t_{i+1}}\int_0^{s'}\chi^{(N)}(x)|\varphi(x,y)|dxdy\right\}
\end{aligned}$$

But $\max_{i\le n-1}\int_{t_i}^{t_{i+1}}\int_0^{s'}\chi^{(N)}(x)|\varphi(x,y)|dxdy\le N$, and converges to zero as $n\to\infty$ almost surely. Hence, $E(\varphi^{(N)}(t))^2=0$ and $\varphi(t,s')=0$ almost surely for any $t\le t'$ and arbitrary z'. Similarly one can show that $\varphi(t',s)=0$ almost surely for any $s\le s'$ and arbitrary z'. Thus,

$$\xi(z)=\xi(0)+\int_{[0,z]}\alpha(u,\xi)du+\int_{[0,z]}\beta(u,\xi)\hat{W}(du).$$ □

2.2 Strong Solution of Stochastic Differentiation Equations

Assume that the stochastic process $\xi(z)$ is given by the equation

$$\xi(dz)=a(z,\xi)dz+b(z,\xi)W(dz), \tag{2.2}$$

where $W(z)$ is a standard Wiener field taking values in R, coefficients $a(z,f)$ and $b(z,f)$ are defined on $z\in[0,T]^2, f\in C[0,T]^2$, and the boundary conditions are given by $\xi(t,0)=\varphi(t)$, $\xi(0,s)=\psi(s)$, $\varphi(0)=\psi(0)=\xi(0)$.

Suppose that the coefficients $a(z,f)$ and $b(z,f)$ in (2.2) are $\mathcal{B}_z$-measurable and linearly bounded, i.e. there exists a nonrandom constant c such that $\|a(\cdot)\|_z+\|b(\cdot)\|_z\le c(1+\|f\|_z)$. (Here $\|\cdot\|$ is a uniform norm in the space of continuous functions $C[0,T]^2$, $\|\cdot\|_z$ is a uniform norm in $C[0,z]$).

Definition 2.3 The function $\xi(z)$, $z\in[0,T]^2$, defined on the probability space $(\Omega,\Im,P)$, is called a strong solution to (2.2) with boundary conditions $\varphi(t)$ and $\psi(s)$, if

(a) $\xi(z)$ is $\Im_z$-adapted and its realizations are continuous with probability 1.
(b) For any z the equality

$$\xi(z)=\varphi(t)+\psi(s)-\xi(0)+\int_{(0,z]}a(u,\xi(u))du+\int_{(0,z]}b(u,\xi(u))W(du)$$

holds with probability 1.

Consider the equation

$$\xi(z)=\varphi(z)+\int_{(0,z]}a(u,\xi(u))du+\int_{(0,z]}b(u,\xi(u))W(du), \tag{2.3}$$

where the stochastic field $\varphi(z)$ is $\Im_z$-adapted and for any z, z', $z \leq z'$, the increment $W(z, z']$ does not depend on the σ-algebra $\Im_z^*$. Let us consider a bit more general equation than (2.2):

$$S(z) = S(\xi, z) = \varphi(z) + \int_{(o,z]} a(u, \xi(\cdot))du + \int_{(o,z]} b(u, \xi(\cdot))W(du).$$

If the stochastic field $\xi(z)$ is continuous and $\Im_z$-adapted, then the field $S(z)$ is defined for any $z \in [0,T]^2$, and its sample functions are continuous. Assume that the stochastic field $\varphi \in \overline{\mathbf{B}}_2$ and $\xi \in \mathbf{B}_2$. Then

$$\|S(\xi,\cdot)\|_z^2 \leq 3\left(\|\varphi(\cdot)\|_z^2 + \int_{(0,z]} c\left(1 + \|\xi(\cdot)\|_u^2\right)du + \left\| \int_{(0,z]} b(u, \xi(\cdot))du \right\|_z^2 \right),$$

and from Theorem 1.17 we obtain

$$E\|S(\xi,\cdot)\|_z^2 \leq 3E\|\varphi(\cdot)\|_z^2 + 51c\left(1 + \int_{(0,z]} E\|\xi(\cdot)\|_u^2 du\right).$$

Similarly, it is easy to obtain from Theorem 1.17 that

$$E\|S(\xi,\cdot) - \varphi(\cdot)\|_z^2 \leq 34c\left(\int_{(0,z]} \left(1 + E\|\xi(\cdot)\|_u^2\right)du\right). \tag{2.4}$$

Definition 2.4 We say that the function $a(z,f)$, where $z \in [0,T]^2$ and $f \in C[0,T]^2$, satisfies the Lipschitz condition with some constant value L, if

$$\|a(\cdot, g(\cdot)) - a(\cdot, h(\cdot))\|_z \leq L\|g(\cdot) - h(\cdot)\|_z$$

for any $z \in [0,T]^2$ and g, $h \in C[0,T]^2$.

Remark 2.1 [23] If the coefficients of (2.3) satisfy the Lipschitz condition, then

$$E\|S(\xi,\cdot) - S(\zeta,\cdot)\|_z^2 \leq 34L^2\left(\int_{(0,z]} \|\xi(\cdot) - \zeta(\cdot)\|_u^2 du\right).$$

Theorem 2.2 *[25, 42, 64, 65, 71] Assume that the functions $a(z,f)$ and $b(z,f)$ are $\mathcal{B}_z$-measurable, satisfy the Lipschitz condition, and the function $\varphi(z) \in \overline{\mathbf{B}}_2$ is continuous. Then (2.3) has a solution $\xi \in \overline{\mathbf{B}}_2$, which is unique in the class $\mathbf{B}_2(\Im_z)$.*

Proof We take an arbitrary function $\xi_0(z) \in \overline{\mathbf{B}}_2(\Im_z)$ and construct the sequence $\xi_n(z)$ by induction: $\xi_{n+1}(z) = S(\xi_n, z)$, $n = 0, 1, 2, \ldots$ Hence, $\xi_n(z) \in \overline{\mathbf{B}}_2(\Im_z)$ and

$$E\|\xi_{n+1}(\cdot)-\xi_n(\cdot)\|_z^2 \le 34L^2\left(\int_{(0,z]}\|\xi_n(\cdot)-\xi_{n-1}(\cdot)\|_u^2 du\right).$$

Put $a := \int_{[0,T]^2} E\|\xi_1(\cdot)-\xi_0(\cdot)\|_u^2 du$. From the previous inequality we derive

$$E\|\xi_{n+1}(\cdot)-\xi_n(\cdot)\|_z^2 \le \frac{a\left(34L^2\right)^n}{(n!)^2} t^n s^n.$$

Let $\beta_n := \|\xi_{n\,+\,1}(\cdot)-\xi_n(\cdot)\|$. Therefore,

$$\sum_{n=1}^{\infty} P\left\{\beta_n > \frac{1}{n^2}\right\} \le \frac{a\left(34L^2\right)^n n^4}{(n!)^2}.$$

By the Borel–Cantelli lemma the series $\sum_{n=0}^{\infty}\beta_n$ converge with probability 1, implying that the series $\sum_{n=0}^{\infty}(\xi_{n+1}(z)-\xi_n(z))$ converge uniformly on $[0,T]^2$ with probability 1. The stochastic field $\xi(z)$ is $\Im_z$-adapted and its sample functions are continuous with probability 1.

Observe that

$$\begin{aligned}
E\|\xi_{n+1}(\cdot)-\xi_n(\cdot)\|^2 &\le \sum_{k=n}^{n+m+1}\frac{1}{k^2}\sum_{k=n}^{n+m-1} Ek^2\|\xi_{k+1}(\cdot)-\xi_k(\cdot)\|^2 \\
&\le \sum_{k=n}^{n+m+1}\frac{1}{k^2}\sum_{k=n}^{n+m-1}\frac{a\left(34L^2\right)^k k^4}{(k!)^4} \to 0.
\end{aligned}$$

Therefore $\xi \in \overline{\mathbf{B}}_2$ and $E\|\xi(\cdot)-\xi_n(\cdot)\|^2 \to 0$ as $n\to\infty$. Recall that inequality (1.8) holds true, and $\xi_{n\,+\,1}(z) = S(\xi_n, z)$. Passing to the limit as $n\to\infty$ we obtain $\xi = S(\xi)$, which proves that the solution of (2.3) exists. Assume now that (2.3) has two solutions in the class $\mathbf{B}_2(\Im_z)$, which we denote, respectively, by η_1 and η_2, and function $\varphi(z)\in\overline{\mathbf{B}}_2$. Let N be a positive variable,

$$I_N(z) = \begin{cases} 1, & \max\left(\|\eta_1\|_z, \|\eta_2\|_z\right) \le N \\ 0, & \text{otherwise.} \end{cases}$$

Then

$$|I_N(\eta_1(z)-\eta_2(z))|^2 \le 2I_N(z)\left|\int_{(0,z]}(a(u,\eta_1(u)-a(u,\eta_2(u))du\right|^2$$
$$+2I_N(z)\left|\int_{(0,z]}(b(u,\eta_1(u)-b(u,\eta_2(u))W(du)\right|^2$$
$$\le 2\left|\int_{(0,z]}I_N(u)(a(u,\eta_1(u)-a(u,\eta_2(u))du\right|^2$$
$$+2\left|\int_{(0,z]}I_N(u)(b(u,\eta_1(u)-b(u,\eta_2(u))W(du)\right|^2,$$

which implies

$$E\|I_N(\eta_1(\cdot)-\eta_2(\cdot))\|_z^2 \le 34L^2\int_{(0,z]}EI_N(u)\|\eta_1(\cdot)-\eta_2(\cdot)\|_u^2 du$$

and, taking into account that $I_N\|\eta_1(\cdot)-\eta_2(\cdot)\| \le 2N$, we obtain

$$E\|I_N(\eta_1(\cdot)-\eta_2(\cdot))\|_z^2 \le 34L^2 4N^2 ts.$$

Thus, for any n

$$E\|I_N(\eta_1(\cdot)-\eta_2(\cdot))\|_z^2 \le 4N^2(34L^2)^n\frac{t^n s^n}{(n!)^2}.$$

Passing to the limit as $n\to\infty$ we obtain that for any $N>0$ $I_N\|\eta_1(\cdot)-\eta_2(\cdot)\|=0$ with probability 1. Thus, $\eta_1(z)=\eta_2(z)$ for any $z\in[0,T]^2$ with probability 1, and therefore, under the conditions of our theorem the solution to (2.2) is unique. □

Remark 2.2 Assume that $\phi(z)=\varphi(t)+\psi(s)-\xi(0)$ and $\varphi(z)\in\overline{\mathbf{B}}_2$. Then the unique solution to (2.3) exists under weaker assumptions, i.e. when the coefficients $a(z,f)$ and $b(z,f)$ satisfy the Lipschitz condition on the subset of $C[0,T]^2$, consisting of the functions with boundary values $\xi(t,0)=\varphi(t)$ and $\xi(0,s)=\psi(s)$.

2.3 Generalized Girsanov Theorem for Stochastic Fields on the Plane

In this section we present results published in [10, 42, 45, 47]. To investigate Ito fields on the plane one often needs to transform the main probability measure P. In some cases it is possible to change the probability measure in such a way that the Ito field transforms into the Wiener field.

Denote

$$\varsigma(z,z',\varphi)=\int_{[z,z']}\varphi(u)W(du)-\frac{1}{2}\int_{[z,z']}\varphi^2(u)du$$

and $\zeta_\varphi(z)=\exp\{\varsigma(0,z,\varphi)\}$, $z\le z'$, $z,\ z'\in[0,T]^2$, where $(\varphi(z),\Im_z)$ is a random field satisfying the condition

$$P\left\{\int_{[0,T]^2}\varphi^2(u)du<\infty\right\}=1. \tag{2.5}$$

It is easy to see that application of the Ito formula (Theorem 1.14) to the function e^x and the field $\zeta(z)$ yields

$$\zeta_\varphi(z)=1+\int_{[0,z]}\zeta_\varphi(t',s)\varphi(t',s')W(dz').$$

Lemma 2.1 *Assume that the stochastic field $(\varphi(z),\Im_z)$ satisfies condition (2.5). Then $E\zeta_\varphi(z)\le 1$, and $E\big(\exp\{\varsigma(z,z',\varphi)\}/\Im_z^*\big)\le 1$ almost surely.*

Proof Put

$$w_N(x)=\begin{cases}1, & |x|\le N,\\ 0, & |x|>N,\end{cases}$$

and

$$\eta_N(z)=w_N\left(\sup_{u\in[0,z]}\zeta_\varphi(u)\right)\zeta(z).$$

Obviously, $\zeta_\varphi(z)=\lim_{N\to\infty}\eta_N(z)$ almost surely. If $\varphi(z)$ is bounded, then $E\eta_N(z)=1$. Passing to the limit as $N\to\infty$ we obtain from the Fatou lemma that $E\zeta(z)\le 1$. Let $\varphi(z)=c(\omega)$, where $c(\omega)$ is the $\Im_z$-measurable random value. Then

$$\begin{aligned}E\big(\exp\{\varsigma(z,z',\varphi)\}/\Im_z^*\big)&=E\left(\exp\left\{c(\omega)W(z,z']-\frac{1}{2}c^2(\omega)(t'-t)(s'-s)\right\}/\Im_z^*\right)\\&=\exp\left\{-\frac{1}{2}c^2(\omega)(t'-t)(s'-s)\right\}\\&\quad\times\exp\left\{\frac{1}{2}c^2(\omega)(t'-t)(s'-s)\right\}=1,\end{aligned}$$

Therefore, the equality $E(\exp\{\varsigma(z,z',\varphi)\}/\Im_z^*) = 1$ holds true for any bounded piecewise constant field $\varphi(z)$. Approximating the stochastic function $\varphi(z)$ by a sequence of bounded piecewise constant fields, we derive in the general case from the Fatou lemma that $E(\exp\{\varsigma(z,z',\varphi)\}/\Im_z^*) \leq 1$. □

Remark 2.3 Under conditions of Lemma 2.1, the stochastic field $(\zeta_\varphi(z), \Im_z)$ is a (nonnegative) super-martingale.

Lemma 2.2 *Assume that all conditions of Lemma 2.1 hold true and*

$$E\exp\left\{\int_{[0,T]^2}\varphi(u)W(du) - \frac{1}{2}\int_{[0,T]^2}\varphi^2(u)du\right\} = 1. \tag{2.6}$$

Then
$E(\exp\{\varsigma(z,z',\varphi)\}/\Im_z^*) = 1$ *almost surely.*

Proof From Lemma 2.1 we obtain

$$\begin{aligned}1 = E\zeta(T,T) \leq E\zeta(z) &= E\exp\{\varsigma(z,z',\varphi) + \varsigma(0,(t,s'),\varphi + \varsigma((t,0),(t',s),\varphi)\}\\ &= E\exp\{\varsigma(0,(t,s'),\varphi + \varsigma((t,0),(t',s),\varphi)\}E(\exp\{\varsigma(z,z',\varphi)\}/\Im_z^*).\end{aligned}$$

Put

$$B_n = \left\{\omega : E(\exp\{\varsigma(z,z',\varphi)\}/\Im_z^*) \leq 1 - \frac{1}{n}\right\}, \quad n = 1,2,\ldots$$

Then

$$\begin{aligned}1 \leq &\int_{\Omega\setminus B_n}\exp\{\varsigma(0,(t,s'),\varphi + \varsigma((t,0),(t',s),\varphi)\}P(d\omega)\\ &+ \left(1 - \frac{1}{n}\right)\int_{B_n}\exp\{\varsigma(0,(t,s'),\varphi + \varsigma((t,0),(t',s),\varphi)\}P(d\omega)\\ &= E\exp\{\varsigma(0,(t,s'),\varphi + \varsigma((t,0),(t',s),\varphi)\}\\ &\quad - \frac{1}{n}\int_{B_n}\exp\{\varsigma(0,(t,s'),\varphi + \varsigma((t,0),(t',s),\varphi)\}P(d\omega).\end{aligned}$$

Further, it follows from Lemma 2.1 that

$$E\exp\{\varsigma(0,(t,s'),\varphi + \varsigma((t,0),(t',s),\varphi)\} \leq 1,$$

implying $P(B_n) = 0$ for any $n > 0$. Therefore,

$$E(\exp\{\varsigma(z,z',\varphi)\}/\Im_z^*) = 1.$$

□

Remark 2.4 Under conditions of Lemma 2.2, the stochastic field $(\zeta_\varphi(z), \Im_z)$ is a martingale.

Since $\zeta(z)$ is a nonnegative martingale, one can define on the probability space $(\Omega, \Im)$ the new probability measure $\widetilde{P}(d\omega) = \exp\{\zeta_\varphi(T,T)\}P(d\omega)$. We denote by $\widetilde{E}$ the expectation with respect to this new measure $\widetilde{P}$.

Lemma 2.3 *Assume that condition (2.6) is fulfilled. Then for any $\Im_{z'}$-measurable random variable $\eta(\omega)$, we have $\widetilde{E}|\eta| < \infty$ with probability 1, and*

$$\widetilde{E}\left(\eta/\Im_z^*\right) = E\left(\eta\exp\{\varsigma(z,z',\varphi\}/\Im_z^*\right), \quad z \leq z' \quad z,z' \in [0,T]^2.$$

Proof It is enough to show that for any bounded $\Im_z^*$-measurable random variable $\gamma(\omega)$ one has

$$\widetilde{E}\left\{\gamma\widetilde{E}\left(\eta/\Im_z^*\right)\right\} = \widetilde{E}\left\{\gamma E\left(\eta\exp\{\varsigma(z,z',\varphi\}/\Im_z^*\right)\right\}.$$

Indeed, $\widetilde{E}\left\{\gamma\widetilde{E}\left(\eta/\Im_z^*\right)\right\} = \widetilde{E}\gamma\eta$. On the other hand, by Lemma 2.2 we obtain

$$\begin{aligned}
&\widetilde{E}\left\{\gamma E\left(\eta\exp\{\varsigma(z,z',\varphi\}/\Im_z^*\right)\right\} = E\gamma E\left(\eta\exp\{\varsigma(z,z',\varphi\}/\Im_z^*\right)\zeta(T,T)\\
&\quad = E\left\{\gamma E\left(\eta\exp\{\varsigma(z,z',\varphi\}/\Im_z^*\right)[\exp\{\varsigma(0,(T,s),\varphi) + \varsigma(0,(t,T),\varphi)\}]\right\}\\
&\quad = E(\gamma\eta\,[\exp\{\varsigma((t,0),(T,s),\varphi) + \varsigma(0,(t,T),\varphi) + \varsigma(z,z',\varphi)\}])\\
&\quad = E(\gamma\eta\,[\exp\{\varsigma((t,0),(T,s),\varphi) + \varsigma(0,(t,T),\varphi) + \varsigma(z,z',\varphi)\}])\\
&\qquad \times E\left[\exp\{\varsigma((t',s),(T,s'),\varphi)\}/\Im_z^*\right]\\
&\quad = E(\gamma\eta\,[\exp\{\varsigma((t,0),(T,s),\varphi) + \varsigma(0,(t,T),\varphi) + \varsigma(z,z',\varphi) + \varsigma((t',s),(T,s'),\varphi)\}])\\
&\qquad \times E\left[\exp\{\varsigma(z',(T,T),\varphi)\}/\Im_z^*\right] = E\gamma\eta\zeta(T,T) = \widetilde{E}\gamma\eta.
\end{aligned}$$
□

Lemma 2.4 *[54] Let $\xi_n \geq 0$, $n = 1, 2, \ldots$ be the sequence of random variables such that $\xi_n \to \xi$ in probability as $n \to \infty$. If $E\xi_n = E\xi = const$, then $\lim_{n\to\infty} E|\xi_n - \xi| = 0$.*

Lemma 2.5 *Assume that $\phi^{(N)}(z)$ is the sequence of $\Im_z$-measurable functions such that $\phi^{(N)}(z) \to \phi(z)$ in probability as $N \to \infty$. If*

$$E\exp\left\{\varsigma\left(z,z',\phi^{(N)}\right)\right\} = E\exp\{\varsigma(z,z',\phi)\} = 1,$$

then

$$\lim_{N\to\infty} E\left|\exp\left\{\varsigma\left(z,z',\phi^{(N)}\right)\right\} - \exp\{\varsigma(z,z',\phi)\}\right| = 0.$$

The proof follows directly from Lemma 2.4.

Theorem 2.3 *(Girsanov theorem). Suppose that the random field* $(\varphi(z), \Im_z)$ *satisfies (2.5) and condition (2.6) is fulfilled. Then the random field* $(\xi(z), \Im_z)$ *given by*

$$\xi(z) = W(z) - \int_{[0,z]} \varphi(u)du,$$

is the Wiener field on the probability space $\left(\Omega, \Im, \widetilde{P}\right)$ *with respect to the flow* $(\Im_z)$ *and probability measure* $\widetilde{P}(d\omega) = Ee^{\zeta(T,T)}P(d\omega)$.

Proof To prove this theorem we construct the sequence of piecewise constant fields $(\varphi_N(z), \Im_z)$, $N = 1, 2, \ldots$, such that

$$\lim_{N\to\infty} \int_{[0,T]^2} (\varphi(u) - \varphi_N(u))^2 du = 0 \text{ a.s.},$$

$$\lim_{N\to\infty} \exp\{\varsigma(z, z', \varphi_N)\} = \exp\{\varsigma(z, z', \varphi)\} \text{ a.s.}$$

From Lemma 1.9,

$$E\exp\{\varsigma(0, (T, T), \varphi_N\} = 1.$$

We prove that for any $N > 0$ the field $(\xi_N(z), \Im_z)$ defined by

$$\xi_N(z) = W(z) - \int_{[0,z]} \varphi_N(u)du,$$

is a Wiener field with respect to the probability measure $\widetilde{P}^N(d\omega) = \exp\{\zeta_{\varphi_N}(T, T)\}$ $P(d\omega)$. Since the field $(\varphi_N(z), \Im_z)$ is bounded,

$$\widetilde{E}_N \exp\{\theta\xi_N(z, z']\} < \infty.$$

Moreover, for any θ

$$\widetilde{E}_N\left(\exp\left\{\theta\xi_N(z, z'] - \frac{\theta^2}{2}(t' - t)(s' - s)\right\} \Big/ \Im_z^*\right)$$
$$= E\left(\exp\left\{\theta\xi_N(z, z'] - \frac{\theta^2}{2}(t' - t)(s' - s) + \varsigma(z, z', \phi_N)\right\} \Big/ \Im_z^*\right)$$
$$= E\left(\exp\left\{\theta W(z, z'] - \theta\int_{[z,z']} \phi_N(u)du - \frac{\theta^2}{2}(t' - t)(s' - s) + \varsigma(z, z', \phi_N)\right\} \Big/ \Im_z^*\right)$$
$$= E\left(\exp\left\{\varsigma(z, z', \phi_N) - \theta\int_{[z,z']} \phi_N(u)du\right\} \Big/ \Im_z^*\right)$$
$$= E\left(\exp\{\varsigma(z, z', \phi_N)(\theta + \phi_N)\}/\Im_z^*\right).$$

The random field $(\varphi_N(z)+\theta, \mathfrak{I}_z)$ is bounded and piecewise constant, which implies by Lemma 2.2

$$E\left(\exp\{\varsigma(z,z',\varphi_N)(\theta+\varphi_N)\}/\mathfrak{I}_z^*\right) = 1.$$

Therefore, by Lemmas 2.3 for any c

$$\begin{aligned}\widetilde{E}_N\left(\exp\{ic\xi_N(z,z'] + \varsigma(z,z',\phi_N)\}/\mathfrak{I}_z^*\right) &= \widetilde{E}_N\left(\exp\{ic\xi_N(z,z']\}/\mathfrak{I}_z^*\right)\\ &= \exp\left\{-\frac{c^2}{2}(t'-t)(s'-s)\right\}.\end{aligned} \tag{2.7}$$

The sequence $\exp\{ic\xi_N(z, z']\}$, $N = 1, 2, \ldots$, is bounded and converges almost surely to $\exp\{ic\xi(z, z']\}$. Note also that $E\exp\{\varsigma(z,z',\varphi_N)\} = 1$. Thus, by Lemma 2.4

$$E\exp\{\varsigma(z,z',\phi)\} = 1,$$

and

$$\lim_{N\to\infty} E\left|\exp\{\varsigma(z,z',\varphi_N)\} - \exp\{\varsigma(z,z',\varphi)\}\right| = 0.$$

Passing to the limit as $N \to \infty$ under expectation in (2.7), we obtain

$$\begin{aligned}\widetilde{E}\left(\exp\{ic\xi(z,z'] + \varsigma(z,z',\varphi)\}/\mathfrak{I}_z^*\right) &= \widetilde{E}\left(\exp\{ic\xi(z,z']\}/\mathfrak{I}_z^*\right)\\ &= \exp\left\{-\frac{c^2}{2}(t'-t)(s'-s)\right\}.\end{aligned}$$

Thus, the random field $(\xi(z), \mathfrak{I}_z)$ is a Wiener field with respect to the probability measure $\widetilde{P}$. □

Definition 2.5 Introduce in the space $C[0,T]^2$ the measure μ_ξ, corresponding to the random field ξ by the following rule: for any $A \in \mathcal{B}$ put $\mu_\xi(A) = P\{\omega : \xi(\omega) \in A\}$.

For some applications it is important to know the sufficient conditions for the existence of μ_ξ and μ_W, as well as the Radon–Nikodym derivatives of these measures.

Theorem 2.4 *Let ξ be a random Ito field of the form*

$$\xi(z) = W(z) + \int_{[0,z]} \varphi(u)du,$$

and let the conditions

$$P\left\{\int_{[0,T]^2}\varphi^2(u)du < \infty\right\} = 1$$

and

$$E\exp\left\{-\int_{[0,T]^2}\varphi(u)W(du) - \frac{1}{2}\int_{[0,T]^2}\varphi^2(u)du\right\} = 1$$

hold true. Then the measure μ_ξ generated by ξ is equivalent to the measure μ_W, corresponding to a standard Wiener field, and

$$\frac{d\mu_W}{d\mu_\xi}(\xi) = E\left[\exp\left\{-\int_{[0,T]^2}\varphi(u)W(du) - \frac{1}{2}\int_{[0,T]^2}\varphi^2(u)du\right\}\Big/\mathfrak{I}^\xi\right] P\text{-a.s.},$$

where $\mathfrak{I}^\xi = \sigma\left\{\xi(z), z\in[0,T]^2\right\}$. If $(\xi(z), \mathfrak{I}_z)$ is a diffusion-type field, then

$$\frac{d\mu_W}{d\mu_\xi}(\xi) = \exp\left\{-\int_{[0,T]^2}\varphi(u,\xi)W(du) + \frac{1}{2}\int_{[0,T]^2}\varphi^2(u,\xi)du\right\},$$

$$\frac{d\mu_\xi}{d\mu_W}(W) = \exp\left\{\int_{[0,T]^2}\varphi(u,W)W(du) - \frac{1}{2}\int_{[0,T]^2}\varphi^2(u,W)du\right\}.$$

Proof Put

$$\zeta(z) = \exp\left\{-\int_{[0,z]^2}\varphi(u)W(du) - \frac{1}{2}\int_{[0,z]^2}\varphi^2(u)du\right\}.$$

From our assumptions, we have $E\zeta(T,T) = 1$, and thus $(\zeta(z), \mathfrak{I}_z)$ is a supermartingale. Denote by $\widetilde{P}$ a measure on the probability space $(\Omega, \mathfrak{I})$ with the property $d\widetilde{P}(\omega) = \zeta(T,T,\omega)dP$. By Theorem 2.3, the stochastic field $(\xi(z), \mathfrak{I}_z)$, $z \in [0,T]^2$ is a Wiener field with respect to the measure $\widetilde{P}$, which implies one has for any $A \in B_T$

$$\mu_W(A) = \widetilde{P}(\xi\in A) = \int_{\{\omega:\xi\in A\}}\zeta(T,T,\omega)dP = \int_{\{\omega:\xi\in A\}}E\left\{\zeta(T,T,\omega)/\mathfrak{I}_T^\xi\right\}dP.$$

Since the random variable $E\left\{\zeta(T,T,\omega)/\mathfrak{I}^\xi_{(T,T)}\right\}$ is $\mathfrak{I}^\xi_{(T,T)}$-measurable, there exists a $\mathcal{B}_{(T,T)}$-measurable nonnegative function $\psi(x)$, such that

$$E\left\{\zeta(T,T,\omega)/\mathfrak{I}^\xi_{(T,T)}\right\} = \psi(\xi(\omega)).$$

Thus,

$$\mu_W(A) = \int_{\{\omega:\xi\in A\}} \psi(\xi(\omega))dP(\omega) = \int_A \psi(x)d\mu_\xi(x).$$

From the last statement we obtain $\mu_W \ll \mu_\xi$ and $\frac{d\mu_W}{d\mu_\xi}(\xi) = \psi(\xi)$ μ_ξ-almost surely, implying $\frac{d\mu_W}{d\mu_\xi}(\xi) = E\left\{\zeta(T,T,\omega)/\Im_T^\xi\right\}$ P-almost surely.

Let us show now that $\mu_\xi \ll \mu_W$. Note that since $P\left\{\int_{[0,T]^2} \varphi(u)W(du) < \infty\right\} = 1$, $\frac{d\widetilde{P}}{dP}(\omega) = \zeta(T,T,\omega)$ and $P\{\zeta(T,T,\omega) = 0\} = 0$. Hence, $\frac{dP}{d\widetilde{P}}(\omega) = \zeta^{-1}(T,T,\omega)$. Therefore, since $\widetilde{P}\{\omega : \xi\in A\} = \mu_W(A)$,

$$\begin{aligned}\mu_\xi(A) = P(\xi\in A) &= \int_{\{\omega:\xi\in A\}} \zeta^{-1}(T,T,\omega)d\widetilde{P}(\omega)\\ &= \int_{\{\omega:\xi\in A\}} \widetilde{E}\left\{\zeta^{-1}(T,T,\omega)/\Im^\xi_{(T,T)}\right\}d\widetilde{P}(\omega)\\ &= \int_A \widetilde{E}\left\{\zeta^{-1}(T,T,\omega)/\Im^\xi_{(T,T)}\right\}_{\xi=x} d\mu_W(x).\end{aligned}$$

As a result, we obtain $\mu_\xi \ll \mu_W$ and

$$\frac{d\mu_\xi}{d\mu_W}(W) = \widetilde{E}\left\{\zeta^{-1}(T,T,\omega)/\Im_T^\xi\right\} P\text{-a.s.}$$ □

Remark 2.5 Usually the assumption

$$E\exp\left\{-\int_{[0,T]^2}\varphi(z)W(dz) - \frac{1}{2}\int_{[0,T]^2}\varphi^2(z)dz\right\} = 1$$

is quite hard to verify. Following [54], it is easy to see that if for any $\delta > 0$

$$E\exp\left\{\left(\frac{1}{2}+\delta\right)\int_{[0,T]^2}\varphi^2(z)dz\right\} < \infty,$$

then the first condition of the Theorem 2.4 is also fulfilled.

Definition 2.6 We say that the stochastic differential equation $\xi(dz) = a(z,\xi)dz + b(z,\xi)W(dz)$ with boundary conditions $\varphi(t)$ and $\psi(s)$ has a weak solution if there exist a probability space $(\Omega, \Im, P)$, a non-decreasing system of σ-algebras $(\Im_z)$, $z \in [0,T]^2$, a continuous random field $(\xi(z), \Im_z)$, and a standard Wiener field $(W(z), \Im_z)$, such that

(a) $\int_{[0,T]^2} |a(z,\xi)|dz < \infty$ P-a.s.

(b) $\int_{[0,T]^2} b^2(z,\xi)dz < \infty$ P-a.s.

(c) For any $z \in [0,T]^2$

$$\xi(z) = \varphi(t) + \psi(s) - \xi(0) + \int_{(o,z]} a(u,\xi(u))du + \int_{(o,z]} b(u,\xi(u))W(du)$$

with probability 1.

In fact, a weak solution is a set of objects $(\Omega, \mathfrak{I}, \mathfrak{I}_z, P, W(z), \xi(z))$.

Definition 2.7 The stochastic differential equation of type (2.2) has a unique weak solution if for any two solutions $(\Omega, \mathfrak{I}, \mathfrak{I}_z, P, W(z), \xi(z))$ and $\left(\widetilde{\Omega}, \widetilde{\mathfrak{I}}, \widetilde{\mathfrak{I}}_z, \widetilde{P}, \widetilde{W}(z), \widetilde{\xi}(z)\right)$ distributions of fields $\xi(z)$ and $\widetilde{\xi}(z)$ coincide, i.e. $\mu_\xi(A) = \widetilde{\mu}_{\widetilde{\xi}}(A)$ for any $A \in \mathcal{B}$.

We find necessary and sufficient conditions for the existence of a unique weak solution in the case $b(z,\xi) \equiv 1$.

Theorem 2.5 *[47] Assume that $a(z,f)$, where $z \in [0,T]^2$ and $f \in C[0,T]^2$, is some measurable functional that does not depend on future, and* $\int_{[0,T]^2} a^2(z,f)dz < \infty$. *Then the equation*

$$\xi(dz) = a(z,\xi)dz + W(dz) \tag{2.8}$$

has a unique weak solution if and only if there exists a Wiener field $(W'(z), \mathfrak{I}'_z)$ on some probability space $(\Omega', \mathfrak{I}', P')$, such that

$$E' \exp\left\{\int_{[0,T]^2} \varphi(u,W')W'(du) - \frac{1}{2}\int_{[0,T]^2} \varphi^2(u,W)du\right\} = 1.$$

Here E' denotes expectation with respect to the measure P'.

Proof Suppose that a weak solution to this equation exists, i.e. for a set of objects $(\Omega, \mathfrak{I}, \mathfrak{I}_z, P, W(z), \xi(z))$, and $z \in [0,T]^2$, we have

$$\xi(z) = \int_{(o,z]} a(u,\xi(u))du + W(z)\, P\text{-a.s.}$$

Note that $\int_{[0,T]^2} a^2(z,\xi)dz < \infty$ and $\int_{[0,T]^2} a^2(z,W)dz < \infty$ P-almost surely, and therefore by Theorem 2.4 the measures corresponding to the fields ξ and W are P-almost surely equivalent, and

$$\frac{d\mu_\xi}{d\mu_W}(W) = \exp\left\{\int_{[0,T]^2} a(u, W(u))W(du) - \frac{1}{2}\int_{[0,T]^2} a^2(u, W(u))du\right\}.$$

Thus, the conditions of the theorem are fulfilled for the standard Wiener field $(W(z), \Im_z)$ from the definition of the weak solution.

Suppose now that the conditions of the theorem hold true. Then the random field

$$\widetilde{W}(z) = W'(z) - \int_{[0,z]} a(u, W'(u))du, \quad z \in [0, T]^2$$

is a Wiener field with respect to the set of σ-algebras $\left(\Im'_z, z \in [0, T]^2\right)$, and the probability measure $\widetilde{P}$ is given by

$$\widetilde{P}(d\omega) = \exp\left\{\int_{[0,T]^2} a(u, W'(u))W'(du) - \frac{1}{2}\int_{[0,T]^2} a^2(u, W'(u))du\right\}P'(d\omega).$$

Therefore, the set of objects $\left(\Omega', \Im', \Im'_z, \widetilde{P}, \widetilde{W}(z), W'(z)\right)$ is a weak solution to (2.8). Moreover, the measure μ_ξ corresponding to any weak solution of (2.8) is equivalent to the measure μ_W. Since the Radon–Nikodym derivative depends only on the functional $a(z,f)$, this solution is unique. □

When dealing with some problems which involve Ito and diffusion-type fields it is convenient to work with a field which admits the following representation:

$$\xi(z) = 1 + \int_{[0,z]} \gamma(t_1, s)W(dz_1), \tag{2.9}$$

where $(\gamma(z), \Im_z)$ is a random field such that

$$P\left\{\int_{[0,T]^2} \gamma^2(u)du < \infty\right\} = 1. \tag{2.10}$$

Lemma 2.6 *Assume that the random field $\xi(z)$ is given by (2.9) and satisfies*

$$P\left\{\xi(z) \geq 0, z \in [0, T]^2\right\} = 1.$$

Then $E\xi(z) \leq 1$.
If in addition

$$E\xi((T, T)) = 1, \tag{2.11}$$

then the random field $(\xi(z), \Im_z)$ is a martingale.

The proof is similar to the proof of Lemma 2.2.
Define on the probability space $(\Omega, \Im)$ *a new probability measure* $\widetilde{P}$ *by*

$$d\widetilde{P}(\omega) = \xi((T,T),\omega)dP(\omega).$$

Lemma 2.7 *Suppose that the stochastic field* ξ *satisfies condition (2.9). Then*

$$\widetilde{P}\left(\inf_{z\in[0,T]^2} \xi(z) = 0\right) = 0.$$

Proof By definition $\widetilde{P}$ we have

$$\widetilde{P}\left(\inf_{z\in[0,T]^2} \xi(z) = 0\right) = \int\limits_{\{\omega:\inf\xi(z)=0\}} \xi(T,T)dP(\omega).$$

Put $D^0 = \{z : \xi(z) = 0\}$. Then $\{\omega : \inf_{z\in D} \xi(z) = 0\} = \{\omega : z\in D^0\}$, and therefore,

$$\widetilde{P}\left(\inf_{z\in[0,T]^2} \xi(z) = 0\right) = \int\limits_{\{\omega:z\in D^0\}} E(\xi(T,T)/\Im_z)dP(\omega) = 0. \qquad \square$$

Lemma 2.8 *Suppose that conditions (2.9) and (2.10) are satisfied, and let* $\alpha = \alpha(\omega)$ *be an* $\Im_z^*$*-measurable random variable with* $E|\alpha| < \infty$. *Then for* $z \in [0,T]^2$, $z \le z'$,

(a) $\widetilde{E}\left(\alpha/\Im_z^2\right) = \overline{\xi}^{-1}(T,s)E\left(\alpha\xi(T,s')/\Im_z^2\right)$.
(b) $\widetilde{E}\left(\alpha/\Im_z^1\right) = \overline{\xi}^{-1}(t,T)E\left(\alpha\xi(t',T)/\Im_z^1\right)$.

Proof Let $\lambda = \lambda(\omega)$ be a bounded $\Im_z^2$-measurable random variable, $0 \le s \le s' \le T$. Then

$$\begin{aligned}\widetilde{E}(\alpha\lambda) &= \widetilde{E}\left(\lambda\widetilde{E}\left(\alpha/\Im_z^2\right)\right) = E\left(\lambda\widetilde{E}\left[\alpha/\Im_z^2\right]\xi(T,T)\right) \\ &= E\left(\lambda\widetilde{E}\left[\alpha/\Im_z^2\right]E\left[\xi(T,T)/\Im_z^2\right]\right) = E\left(\lambda\widetilde{E}\left[\alpha/\Im_z^2\right]\xi(T,s)\right).\end{aligned}$$

On the other hand,

$$\begin{aligned}\widetilde{E}(\alpha\lambda) &= E(\lambda\alpha\xi(T,T)) = E\left(\lambda\alpha E\left[\xi(T,T)/\Im_z^2\right]\right) \\ &= E(\alpha\lambda\xi(T,s')) = E\left(\lambda\widetilde{E}\left[\alpha\xi(T,s')/\Im_z^2\right]\right).\end{aligned}$$

Therefore, $\xi(T,s)\widetilde{E}\left[\alpha/\Im_z^2\right] = E\left(\alpha\xi(T,s')/\Im_z^2\right)$ with P- and $\widetilde{P}$ -probability 1. Since $\widetilde{P}\{\xi(z) > 0\} = 1$, we have $\widetilde{P}\left\{\overline{\xi}^{-1}(z) = \xi^{-1}(z)\right\} = 1$, and

$$\widetilde{E}\left(\alpha/\Im_z^2\right) = \overline{\xi}^{-1}(T,s)E\left(\alpha\xi(T,s')/\Im_z^2\right)$$

for all $z \le z'$, which proves the assertion (a) The proof of (b) is similar. □

Theorem 2.6 *(generalized Girsanov theorem) Let $(\xi(z), \Im_z)$ be a random field of the form (2.9) and assume that conditions (2.10) and (2.11) are fulfilled. Then the random field $\widetilde{W} = \left(\widetilde{W}(z), \Im_z\right)$, z $\in [0,T]^2$, given by*

$$\widetilde{W}(z) = W(z) - \int_{[0,z]} \overline{\xi}^{-1}(t', s)\gamma(z')dz' \tag{2.12}$$

is a Wiener field on the probability space $\left(\Omega, \Im, \widetilde{P}\right)$ with respect to the set of σ-algebras $\Im_z$ and the measure $\widetilde{P}$.

Proof Since $\widetilde{P}\{\xi(z) = 0\} = 0$ and $\widetilde{P}\left\{\overline{\xi}^{-1}(z) = \xi^{-1}(z)\right\} = 1$, the field $\overline{\xi}^{-1} = \left(\overline{\xi}^{-1}(z), \Im_z\right)$, $z \in [0,T]^2$ has continuous sample paths and, therefore,

$$\widetilde{P}\left\{\sup_{z\in[0,T]^2} \overline{\xi}^{-1}(z) < \infty\right\} = 1.$$

Moreover, the measure $\widetilde{P}$ is absolutely continuous with respect to the measure P, and

$$\widetilde{P}\left\{\int_{[0,T]^2} \gamma^2(u)du < \infty\right\} = 1.$$

Observe that

$$\int_{[0,T]^2}\left(\overline{\xi}^{-1}(z)\gamma(z)\right)^2 dz \le \sup_{z\in[0,T]^2}\left(\overline{\xi}^{-1}(z)\right)^2\int_{[0,T]^2}\left(\overline{\xi}^{-1}(z)\gamma(z)\right)^2 dz.$$

Hence, the integral in (2.12) is well defined. To prove the theorem, it suffices to show that for any $z = (t,s) \in [0,T]^2$ the random processes $\left\{\widetilde{W}(x,s), \Im^1_{(x,s)},\ 0 \le x \le T\right\}$ and $\left\{\widetilde{W}(t,y), \Im^2_{(t,y)},\ 0 \le y \le T\right\}$ are one-parameter Wiener processes with parameters s and t, respectively.

Fix t, $0 \leq t \leq T$. Assume that for some constant values c_1 and c_2 we have

$$P\left(0 < c_1 \leq \inf_{z \in D} \xi(z) \leq \sup_{z \in D} \xi(z) \leq c_2 < \infty\right) = 1, \tag{2.13}$$

and

$$E\left(\int_{[0,T]^2} \gamma^2(u)du\right) < \infty. \tag{2.14}$$

Put

$$\varsigma(t, y, s) = \exp\left\{i\theta\left(\widetilde{W}(t, y) - \widetilde{W}(t, s)\right)\right\}.$$

Lemma (1.5) implies that

$$E\left[\varsigma/\Im_z^2\right] = \overline{\xi}^{-1}(T, s)E\left[\varsigma\xi(T, y)/\Im_z^2\right]\widetilde{P} - \text{a.s.}$$

Therefore, using the Ito formula (Theorem 1.16), we get

$$\begin{aligned}&\varsigma(t, y, s)\,\xi(T, y) - \varsigma(t, s, s)\,\xi(T, s)\\&= \int_0^t \int_s^y \varsigma(t, v, s)\xi(T, v)\overline{\xi}^{-1}(t, s)\gamma(u, v)W(du, dv).\end{aligned}$$

Conditions (2.13) and (2.14) ensure the existence of all integrals in the above equation. Taking the conditional expectation with respect to the σ-algebra $\Im_z^2$ we see that

$$\overline{\xi}^{-1}(T, s)E\left[\varsigma\xi(T, y)/\Im_z^2\right] = 1 - \frac{\theta^2}{2}\int_0^t \int_s^y \overline{\xi}^{-1}(T, s)E\left(\varsigma(t, v, s)\xi(T, v)/\Im_z^2\right)du, dv$$

with P- and $\widetilde{P}$-probability 1, whence

$$\overline{\xi}^{-1}(T, s)E\left[\varsigma\xi(T, y)/\Im_z^2\right] = \exp\left\{-\frac{\theta^2}{2}t(y - s)\right\}.$$

Therefore,

$$E\exp\left\{i\theta\left(\widetilde{W}(t, y) - \widetilde{W}(t, s)\right)\right\} = \exp\left\{-\frac{\theta^2}{2}t(y - s)\right\}$$

$\widetilde{P}$-almost surely.

Now suppose that conditions (2.13) and (2.14) are violated. For every integer n introduce the sets

$$D^n = \left\{ z\in[0,T]^2 : \int_{[0,z]} \gamma^2(t',s)dz' + \left(\inf_{z'\leq z}\xi(z')\right)^{-1} + \sup_{z'\leq z}\xi(z') \leq n \right\},$$

which form the sequence of stopping regions $D^n \subset D^{n+1} \subset \ldots$. Since

$$P\left\{ \int_{[0,T]^2} \gamma^2(t',s)dz' + \sup_{z\in[0,T]^2} \xi(z) \leq \infty \right\} = 1$$

and

$$\widetilde{P}\left\{ \inf_{z\in[0,T]^2} \xi(z) > 0 \right\} = 1, \tag{2.15}$$

and the measure $\widetilde{P}$ is absolutely continuous with respect to P, we conclude that $D^n \uparrow [0,T]^2$ as $n \to \infty$ $\widetilde{P}$-almost surely.

Put $\gamma_n(z) = \gamma(z)\chi_{z\in D^n}$,

$$\xi_n(z) = 1 + \int_{[0,z]} \gamma_n(t',s)W(dz'),$$

$$\widetilde{W}_n(z) = W(z) - \int_{[0,z]} \overline{\xi}_n^{-1}(t',s)\gamma_n(t',s)dz'.$$

The random field $\xi_n = (\xi_n(z), \Im_z)$, $z \in [0,T]^2$, is a martingale, $E\xi_n(T,T) = 1$, and satisfies conditions (2.12) and (2.13) with $c_1 = 1/n$ and $c_2 = n$. Define the probability measure $\widetilde{P}_n$ by the equality $d\widetilde{P}_n(\omega) = \xi_n(T,T)P(\omega)$. From above,

$$\widetilde{E}_n\left[\exp\left\{i\theta\left(\widetilde{W}_n(t,y) - \widetilde{W}_n(t,s)\right)\right\}/\Im_z^2\right] = \exp\left\{-\frac{\theta^2}{2}t(y-s)\right\},$$

where $0 \leq s \leq y \leq T$ and $\widetilde{E}_n$ is the expectation with respect to the measure $\widetilde{P}_n$. Since

$$\widetilde{E}_n\left[\exp\left\{i\theta\left(\widetilde{W}_n(t,y) - \widetilde{W}_n(t,s)\right)\right\}/\Im_z^2\right] \to \widetilde{E}\left[\exp\left\{i\theta\left(\widetilde{W}(t,y) - \widetilde{W}(t,s)\right)\right\}/\Im_z^2\right]$$

in $\widetilde{P}$ a.s. $n \to \infty$, it suffices to verify that

$$\overline{\lim_{n\to\infty}}\, \widetilde{E}\left|\widetilde{E}_n\left[\exp\left\{i\theta\left(\widetilde{W}_n(t,y)-\widetilde{W}_n(t,s)\right)\right\}/\mathfrak{I}_z^2\right]\right.$$
$$\left.-\widetilde{E}\left[\exp\left\{i\theta\left(\widetilde{W}_n(t,y)-\widetilde{W}_n(t,s)\right)\right\}/\mathfrak{I}_z^2\right]\right|=0.$$

For each n the measure $\widetilde{P}_n$ is equivalent to the measure P, that is, $\dfrac{d\widetilde{P}_n}{dP}=\dfrac{1}{\xi_n(T,T)}$. Applying Lemma 2.8 and performing simple transformations, we see that

$$\widetilde{E}_n\left[\exp\left\{i\theta\left(\widetilde{W}_n(t,y)-\widetilde{W}_n(t,s)\right)\right\}/\mathfrak{I}_z^2\right]$$
$$=E\left[\exp\left\{i\theta\left(\widetilde{W}_n(t,y)-\widetilde{W}_n(t,s)\right)\right\}\xi(T,y)\overline{\xi}^{-1}(T,s)\right],$$

$$\widetilde{E}_n\left[\exp\left\{i\theta\left(\widetilde{W}_n(t,y)-\widetilde{W}_n(t,s)\right)\right\}/\mathfrak{I}_z^2\right]$$
$$=E\left[\exp\left\{i\theta\left(\widetilde{W}_n(t,y)-\widetilde{W}_n(t,s)\right)\right\}\frac{\xi_n(T,y)}{\xi_n(T,s)}\right],$$

with $\widetilde{P}$-probability 1. Therefore,

$$\widetilde{E}\left|\widetilde{E}_n\left[\exp\left\{i\theta\left(\widetilde{W}_n(t,y)-\widetilde{W}_n(t,s)\right)\right\}/\mathfrak{I}_z^2\right]\right.$$
$$\left.-\widetilde{E}\left[\exp\left\{i\theta\left(\widetilde{W}_n(t,y)-\widetilde{W}_n(t,s)\right)\right\}/\mathfrak{I}_z^2\right]\right|$$
$$\leq\widetilde{E}\left|E\left[\exp\left\{i\theta\left(\widetilde{W}_n(t,y)-\widetilde{W}_n(t,s)\right)\right\}\frac{\xi_n(T,y)}{\xi_n(T,s)}\right].\right.$$
$$\left.-\left[\exp\left\{i\theta\left(\widetilde{W}_n(t,y)-\widetilde{W}_n(t,s)\right)\right\}\xi(T,y)\overline{\xi}^{-1}(T,s)\right]\right|$$
$$\leq\widetilde{E}E\left[\left|\frac{\xi_n(T,y)}{\xi_n(T,s)}-\xi(T,y)\overline{\xi}^{-1}(T,s)\right|/\mathfrak{I}_z^2\right]$$
$$=E\left|\xi(T,s)\,\xi_n(T,y)\,\overline{\xi}_n^{-1}(T,s)-\xi(T,s)\,\xi(T,y)\,\xi^*(T,s)\right|.$$

Introduce the stopping region $D^0=\left\{z\in[0,T]^2:\ \inf_{z'\leq z}\xi(T,s')>0\right\}$. It follows from condition (2.15) that the above-defined stopping regions D^n converge to D^0 as $n\to\infty$ with P-probability 1. Thus, $\xi(T,s)=\xi_0(T,s)$ for all $0\leq s\leq T$, where $\xi_0(z)$ is defined in the same way as $\xi_n(z)$.

Now we shall to verify that

$$\overline{\lim_{n\to\infty}}\,\xi_0(T,s)\,\xi_n(T,y)\,\overline{\xi}_n^{-1}(T,s)=\xi_0(T,s)\,\xi_0(T,y)\,\overline{\xi}_0^{-1}(T,s). \tag{2.16}$$

By continuity of $\xi(z)$, this equality is valid provided that

$$\xi_0(T,s)\overline{\xi}_n^{-1}(T,s) = \xi_0(T,s)\,\overline{\xi}_0^{-1}(T,s) P\text{-a.s.}$$

If the point (T,s) does not belong to D^0, then $\xi_0(T,s)\ \xi_n^+(T,s) = 0$ for all n. Since on the set D^0 we have $\inf_n \xi_n(T,s) > 0$, the equality (2.16) holds true. Hence,

$$\xi(T,s)\,\xi_n(T,y)\overline{\xi}_n^{-1}(T,s) \to \xi(T,s)\,\xi(T,y)\overline{\xi}^{-1}(T,s), \quad n\to\infty P\text{-a.s.}$$

Next,

$$E\Big(\xi(T,s)\overline{\xi}^{-1}(T,s)\xi(T,y)\Big) = E\Big(\xi(T,s)\overline{\xi}^{-1}(T,s)E\big[\xi(T,y)/\Im_{(T,s)}\big]\Big) = 1,$$

$$E\big(\xi(T,s)\,\xi_n(T,y)\,\xi_n^{-1}(T,s)\big) = E\big(\xi(T,s)\ \xi_n^{-1}(T,s)E\left[\xi_n(T,y)/\Im_{(T,s)}\right]\big) = 1.$$

Thus,

$$\overline{\lim_{n\to\infty}}\ E\Big|\xi_0(T,s)\,\xi_n(T,y)\overline{\xi}_n^{-1}(T,s) - \xi(T,s)\,\xi(T,y)\overline{\xi}^{-1}(T,s)\Big| = 0,$$

which implies the required assertion. □

To summarize, we have shown that under the conditions of the theorem $\left(\widetilde{W}(z),\Im_z^2\right), 0\le t\le T$, is a one-parameter martingale for any fixed s. The assertion that $\left(\widetilde{W}(z),\Im_z^1\right)$ is a one-parameter Wiener $\widetilde{P}$-martingale with parameter t can be proved in the same fashion by applying part (b) of Lemma 2.8.

2.4 Some Properties of Measures Corresponding to Random Fields on the Plane

In this section we present results published in [12, 14, 16, 42, 47, 48]. Let $(\xi(z),\Im_z)$ be a continuous random diffusion-type field, given by

$$\xi(z) = \int_{[0,z]} \varphi(u,\xi)du + W(z), \tag{2.17}$$

for any $z \in [0,T]^2$, where $\varphi(z,f), f \in C[0,T]^2$ is some measurable functional independent of the future, and the condition

$$P\left\{\int_{[0,T]^2} |\varphi(u,\xi)|du < \infty\right\} = 1$$

holds true. In this section we study the properties of the random field $\zeta(z, W) = \frac{d\mu_\xi}{d\mu_W}(z, W)$.

Define $\mathfrak{I}_z^W := \sigma\{\omega : \ W(z'), \ z' \le z\}$, where the σ-algebras $\mathfrak{I}_z^W$ are completed by the sets from initial σ-algebra $\mathfrak{F}$ with P-measure 0.

It is easy to see that the field $(\zeta(z, W), \mathfrak{I}_z^W)$ is a martingale. Indeed, assume that $z < z'$ and that $\lambda(W)$ is some bounded $\mathfrak{I}_z^W$-measurable random variable. Then

$$E\lambda(W)\zeta(z', W) = \int \lambda(x)\frac{d\mu_\xi}{d\mu_W}(z', x)d\mu_W(x) = \int \lambda(x)d\mu_\xi(z', x)$$
$$= \int \lambda(x)\zeta(z, x)d\mu_W(z, x),$$

implying $E(\zeta(z, z'](W)/\mathfrak{I}_z^W) = 0$. Further we assume that the field $(\zeta(z, W), \mathfrak{I}_z^W)$ satisfies the stronger condition, namely, that it is a strong martingale:

$$E(\zeta(z, z'](W)/\mathfrak{I}_z^{W*}) = 0. \tag{2.18}$$

Theorem 2.7 *Let* $(\xi(z), \mathfrak{I}_z)$, $z \in [0,T]^2$, $\xi(0) = 0$, *be a diffusion-type random field (2.17). Assume that condition (2.18) is fulfilled, and*

$$P\left\{\int_{[0,T]^2} \varphi^2(u, \xi)du < \infty\right\} = 1.$$

Then the field $\zeta(z,W)$, $z \in [0,T]^2$ *is the unique solution to the equation*

$$\zeta(z, W) = 1 + \int_{[0,z]} \zeta(t', s)\varphi(z')W(dz'). \tag{2.19}$$

Proof It follows that $E\zeta((T,T),W) = 1$ from the definition of $\zeta(z,W)$. Then, condition (2.18) allows to apply Theorem 2.6 (the Girsanov theorem) to $\zeta(z,W)$. By this theorem the random field $\zeta(z,W)$ is continuous with probability 1 and can be represented as

$$\zeta(z, W) = 1 + \int_{[0,z]} \gamma(t_1, s) W(dz_1),$$

where $\gamma(z)$ is a $\mathfrak{I}_z$-measurable stochastic field with

$$E\left(\int_{[0,T]^2} \gamma^2(z)dz\right) < \infty.$$

Therefore, we can introduce a new probability measure $\widetilde{P}$ by $d\widetilde{P}(\omega) = \zeta((T,T),W)dP(\omega)$. We consider on the probability space $\left(\Omega,\Im,\widetilde{P}\right)$ a random field $\widetilde{W} = \left(\widetilde{W}(z),\Im_z^W\right)$, $z \in [0,T]^2$,

$$\widetilde{W}(z) = W(z) - \int_{[0,z]} B(z',\omega)dz',$$

$$B(z',\omega) = \overline{\zeta}^{-1}(t',s)\gamma(t',s').$$

It is a Wiener random field, and

$$\widetilde{P}\left\{\int_{[0,T]^2} B^2(z)dz < \infty\right\} = 1.$$

Thus, by Theorem 1.14 there exists a functional $\beta(z,f)$, where $f \in C[0,T]^2$, such that for almost all $z \in [0,T]^2$ we have $B(z) = \beta(z,W)$ and

$$\widetilde{W}(z) = W(z) - \int_{[0,z]} \beta(z',W)dz'$$

with probability 1. Since the measure μ_ξ is absolutely continuous with respect to the measure μ_W, we have

$$\begin{aligned}
P&\left\{\int_{[0,T]^2} \beta^2(z,\xi)\,dz < \infty\right\} = \mu_\xi\left\{x : \int_{[0,T]^2} \beta^2(z,x)\,dz < \infty\right\} \\
&= \int \chi_{\int_{[0,T]^2} \beta^2(z,x)dz < \infty}(x)d\mu_\xi(x) = \int \chi_{\int_{[0,T]^2} \beta^2(z,x)dz < \infty}(x)\zeta((T,T),x)d\mu_W(x) \\
&= \widetilde{P}\left\{\int_{[0,T]^2} \beta^2(z,\xi)dz < \infty\right\} = 1
\end{aligned}$$

Define on the probability space $(\Omega,\Im,P)$ a random field $\hat{W} = \left(\hat{W}(z),\Im_z^\xi\right)$, $z \in [0,T]^2$, by

$$\hat{W}(z,f) = f(z) - \int_{[0,z]} \beta(z',f)dz',$$

where $f \in C[0,T]^2$. If $f = \xi$, then $\hat{W} = \left(\hat{W}(z),\Im_z^\xi\right)$ is a Wiener field. Indeed, let $\lambda(\xi)$ be a bounded random $\Im_z^\xi$-measurable variable. Then for any $z' > z$

$$E\lambda(\xi)\exp\{i\theta\hat{W}(z,z']\} = E(\lambda(\xi)E[\exp\{i\theta\hat{W}(z,z']\}/\Im_z^\xi]).$$

On the other hand,

$$\begin{aligned}
E\lambda(\xi)\exp\{i\theta\hat{W}(z,z']\} &= \int\lambda(f)\exp\{i\theta f(z,z']\}d\mu_\xi(f) \\
&= \int\lambda(f)\exp\{i\theta f(z,z']\}\zeta((T,T),f)d\mu_W(f) \\
&= \int\lambda(W)\exp\left\{i\theta\widetilde{W}(z,z']\right\}\zeta((T,T),W)dP \\
&= \widetilde{E}\,\lambda(W)\exp\left\{i\theta\widetilde{W}(z,z']\right\} \\
&= \widetilde{E}\,\lambda(W)\exp\left\{-\frac{\theta^2}{2}(t'-t)(s'-s)\right\} \\
&= \exp\left\{-\frac{\theta^2}{2}(t'-t)(s'-s)\right\}E\lambda(\xi),
\end{aligned}$$

whence

$$E\left(\lambda(\xi)E\left[\exp\{i\theta\hat{W}(z,z']\}/\mathfrak{I}_z^\xi\right]\right) = \exp\left\{-\frac{\theta^2}{2}(t'-t)(s'-s)\right\}$$

as $E\left[\lambda(\xi)/\mathfrak{I}_z^\xi\right] = 1$. Therefore,

$$\hat{W}(z) - W(z) = \int_{[0,z]}(\varphi(z',\xi) - \beta(z',\xi))dz',$$

where $\left(\hat{W}(z),\mathfrak{I}_z^\xi\right)$ and $\left(W(z),\mathfrak{I}_z^\xi\right)$ are two Wiener fields. This means that $\hat{W}(z)$ $-W(z) = 0$ with probability 1. Hence, for almost all $z \in [0,T]^2$ we have $P\{\varphi(z,\xi)$ $= \beta(z,\xi)\} = 1$, and taking into account that $P\{\zeta(z,\xi) = 0\} = 0$, we get

$$\zeta(z,W)\varphi(z,W) = \zeta(z,W)\beta(z,W).$$

for almost all $z \in [0,T]^2$ with probability 1.

By definition, $\beta(z',\omega) = \overline{\zeta}^{-1}(t',s,W)\gamma(t',s',W)$, whence

$$\begin{aligned}
&P\left\{\int_{[0,T]^2}(\zeta(t',T,W)\varphi(z',W))^2dz' < \infty\right\} \\
&= P\left\{\int_{[0,T]^2}\left(\zeta(t',T,W)\overline{\zeta}^{-1}(t',T,W)\gamma(z')\right)^2dz' < \infty\right\} \\
&\geq P\left\{\int_{[0,T]^2}\gamma^2(z)dz < \infty\right\} = 1
\end{aligned}$$

Therefore, the stochastic integral $\int_{[0,z]}\zeta(t',s,W)\varphi(z')W(dz')$ is well defined. Let $D = \{z \in [0,T]^2 : \zeta(z) \neq 0\}$. We have

$$\zeta(z) = E\zeta(z) + \int_{[0,z]} \gamma(t_1,s)W(dz_1),$$

and by definition

$$1 + \int_{[0,z]} \zeta(t',s,W)\phi(z',W)W(dz') = 1 + \int_{[0,z]} \zeta(t',s,W)\overline{\zeta}^{-1}(t',s,W)\gamma(z')W(dz').$$

Moreover, $\zeta(z) = 0$ if $z \notin D$. Thus, for all $z \in D$

$$1 + \int_{[0,z]} \zeta(t',s,W)\varphi(z',W)W(dz') = 1 + \int_{[0,z]} \gamma(t_1,s)W(dz_1),$$

and by (2.19) we obtain the required assertion. □

Theorem 2.8 *Suppose that the following conditions hold true:*

$$P\left\{\int_{[0,T]^2} \varphi^2(u,\xi)du < \infty\right\} = 1, \tag{2.20}$$

$$P\left\{\int_{[0,T]^2} \varphi^2(u,W)du < \infty\right\} = 1. \tag{2.21}$$

Then the measures μ_ξ and μ_W are equivalent, with Radon–Nikodym derivatives given by

$$\frac{d\mu_\xi}{d\mu_W}(W) = \exp\left\{\int_{[0,T]^2} \varphi(u,W(u))W(du) - \frac{1}{2}\int_{[0,T]^2} \varphi^2(u,W(u))du\right\}, \tag{2.22}$$

$$\frac{d\mu_W}{d\mu_\xi}(\xi) = \exp\left\{-\int_{[0,T]^2} \varphi(u,\xi(u))\xi(du) - \frac{1}{2}\int_{[0,T]^2} \varphi^2(u,\xi(u))du\right\}, \tag{2.23}$$

where the equalities are satisfied with probability 1.

Proof Denote

$$\chi^{(n)}(z,x) = \chi_{\left\{x:\, x\in C,\ \int_{[0,z]} \varphi^2(z_1,x)dz_1 < n\right\}},$$

$$\varphi^{(n)}(z,x) = \varphi(z,x)\chi^{(n)}(z,x),$$

$$\xi^{(n)}(z) = \int_{[0,z]} \varphi^{(n)}(z_1,\xi)dz_1 + W(z), \quad z\in[0,T]^2, \quad n = 1,2,\dots$$

It is easy to see that $\varphi^{(n)}(z,\xi) = \varphi^{(n)}(z,\xi^{(n)})$, $z \in [0,T]^2$, P-almost surely. Thus the field $\left(\xi^{(n)}(z), \mathfrak{I}_z\right)$ is a diffusion-type field, and

$$\xi^{(n)}(z) = \int_{[0,z]} \varphi^{(n)}\left(z_1,\xi^{(n)}\right)dz_1 + W(z), \quad z\in[0,T]^2, P\text{-a.s.}$$

Since

$$P\left\{\int_{[0,T]^2}\left(\varphi^{(n)}\left(z,\xi^{(n)}\right)\right)^2 dz < n\right\} = 1,$$

then, taking into account Remark 2.5, we obtain

$$E\exp\left\{-\int_{[0,T]^2}\varphi^{(n)}\left(z,\xi^{(n)}\right)W(dz) - \frac{1}{2}\int_{[0,T]^2}\left(\varphi^{(n)}\left(z,\xi^{(n)}\right)\right)^2 dz\right\} = 1,$$

which gives by Theorem 2.4 the equivalence $\mu_{\xi^{(n)}} \sim \mu_W$, where

$$\frac{d\mu_W}{d\mu_{\xi^{(n)}}}\left(\xi^{(n)}\right) = \exp\left\{-\int_{[0,T]^2}\varphi^{(n)}\left(z,\xi^{(n)}\right)W(dz) + \frac{1}{2}\int_{[0,T]^2}\left(\varphi^{(n)}\left(z,\xi^{(n)}\right)\right)^2 dz\right\} P\text{-a.s.}$$

The condition (2.20) guarantees that the property

$$\lim_{n\to\infty}\frac{d\mu_W}{d\mu_{\xi^{(n)}}}\left(\xi^{(n)}\right) = \rho(\xi)$$

holds true P-a.s., where

$$\rho(\xi) = \exp\left\{-\int_{[0,T]^2}\varphi(z,\xi)W(dz) + \frac{1}{2}\int_{[0,T]^2}\varphi^2(z,\xi)dz\right\}.$$

Let us show that the set of random variables $\left\{\frac{d\mu_W}{d\mu_{\xi^{(n)}}}\left(\xi^{(n)}\right),\ n = 1,2,\dots\right\}$ is uniformly integrable. Indeed,

$$\int_{\frac{d\mu_W}{d\mu_{\xi^{(n)}}}\left(\xi^{(n)}\right)>N}\frac{d\mu_W}{d\mu_{\xi^{(n)}}}\left(\xi^{(n)}\right)P(d\omega)=\int_{\frac{d\mu_W}{d\mu_{\xi^{(n)}}}\left(\xi^{(n)}\right)>N}\frac{d\mu_W}{d\mu_{\xi^{(n)}}}(z)\mu_{\xi^{(n)}}(dz)$$

$$=\mu_W\left\{z\ :\ \frac{d\mu_W}{d\mu_{\xi^{(n)}}}(z)>N\right\}=P\left\{z\ :\ \frac{d\mu_W}{d\mu_{\xi^{(n)}}}(W)>N\right\}$$

$$=P\left\{-\int_{[0,T]^2}\phi^{(n)}(z,W)W(dz)+\frac{1}{2}\int_{[0,T]^2}\left(\phi^{(n)}(z,W)\right)^2dz>\ln N\right\}$$

$$\leq P\left\{\left|\int_{[0,T]^2}\phi^{(n)}(z,W)W(dz)\right|>\frac{\ln N}{2}\right\}$$

$$+P\left\{\int_{[0,T]^2}\left(\phi^{(n)}(z,W)\right)^2dz>\frac{\ln N}{2}\right\}$$

$$\leq\frac{4}{\ln N}+2P\left\{\int_{[0,T]^2}\phi^2(z,W)dz>\ln N\right\},\quad N=1,2,\ldots$$

From condition (2.20) we get

$$\sup_n\int_{\frac{d\mu_W}{d\mu_{\xi^{(n)}}}\left(\xi^{(n)}\right)>N}\frac{d\mu_W}{d\mu_{\xi^{(n)}}}\left(\xi^{(n)}\right)P(d\omega)\to 0,\quad N\to\infty.$$

Thus,

$$E\exp\left\{-\int_{[0,T]^2}\varphi(z,\xi)\xi(dz)+\frac{1}{2}\int_{[0,T]^2}\varphi^2(z,\xi)dz\right\}=1$$

and conditions (2.22) and (2.23) are satisfied. Theorem is proved. □

Theorem 2.9 *Let* $(\xi(z),\Im_z)$, $z\in[0,T]^2$ *be the generalized Ito field of the form*

$$\xi(z)=\int_{[0,z]}\alpha(z')dz'+\int_{[0,z]}\beta(z')W(dz')$$

with coefficients satisfying

$$P\left\{\int_{[0,T]^2}|\alpha(z)|dz<\infty\right\}=1,\tag{2.24}$$

$$P\left\{\int_{[0,T]^2}\beta^2(z)dz<\infty\right\}=1.\tag{2.25}$$

Let $\widetilde{P}(d\omega) = \exp\{\zeta(0,T,\varphi\}P(d\omega)$, *hence* $\varphi(z)$ *satisfies*

$$P\left\{\int_{[0,T]^2} \varphi^2(z)dz < \infty\right\} = 1 \quad \text{and} \quad E\exp\{\zeta(0,T,\varphi)\} = 1.$$

Then

1. $\left(\xi(z), \Im_z, \widetilde{P}\right)$ is an Ito field with coefficients $\widetilde{\alpha}(z) = \alpha(z) + \beta(z)\varphi(z)$ and $\widetilde{\beta}(z) = E(\beta(z)/\Im_z^W)$

$$\widetilde{\xi}(z) = W(z) - \int_{[0,z]} \varphi(z')dz'.$$

2. $\mu_\xi \ll \mu_{\widetilde{\xi}}$.

Proof It is easy to see that for the random field $\psi = (\psi(z), \Im_z)$ such that

$$P\left\{\int_{[0,T]^2} \psi^2(z)dz < \infty\right\} = 1$$

both stochastic integrals $\int_{[0,z]}\psi(z')W(dz')$ and $\int_{[0,z]} \psi(z')\widetilde{\xi}(dz')$ exist and, moreover,

$$\int_{[0,z]} \psi(z')W(dz') = \int_{[0,z]} \psi(z')\widetilde{\xi}(dz') + \int_{[0,z]} \psi(z')\varphi(z')dz',$$

implying that

$$\begin{aligned}\xi(z) &= \int_{[0,z]} \alpha(z')dz' + \int_{[0,z]} \beta(z')W(dz') \\ &= \int_{[0,z]} [\alpha(z') + \beta(z')\varphi(z')]dz' + \int_{[0,z]} \beta(z')\widetilde{\xi}(dz').\end{aligned}$$

Thus, statement (1) is proved. Statement (2) follows directly from the relation $\widetilde{P} \ll P$. □

Theorem 2.10 *Let* $(\xi(z), \Im_z)$, $z \in [0,T]^2$ *be a generalized Ito field of the form*

$$\xi(z) = \int_{[0,z]} A(z')dz' + \int_{[0,z]} b(z',\xi)W(dz'),$$

and let $(\eta(z), \Im_z)$ *be a generalized diffusion-type field,*

$$\eta(z) = \int_{[0,z]} a(z',\eta)dz' + \int_{[0,z]} b(z',\eta)W(dz'). \tag{2.26}$$

Assume that the following conditions hold true:

(a) Functionals $a(z,x)$ and $b(z,x)$ are such that there exists a unique strong solution to (2.26).
(b) For any $z \in [0,T]^2$ the equation $b(z,\xi)\alpha(z) = A(z) - a(z,\xi)$ has a bounded solution with respect to $\alpha(z)$.
(c) $P\left\{\int_{[0,T]^2} \alpha^2(z)dz < \infty\right\} = 1.$
(d) $E\exp\left\{-\int_{[0,T]^2} \alpha(z)W(dz) - \frac{1}{2}\int_{[0,T]^2} \alpha^2(z)dz\right\} = 1.$

Then $\mu_\xi \sim \mu_\eta$, and

$$\frac{d\mu_\eta}{d\mu_\xi}(\xi) = E\left[\exp\left\{-\int_{[0,T]^2} \alpha(z)W(dz) - \frac{1}{2}\int_{[0,T]^2} \alpha^2(z)dz\right\}/\Im^{\xi}_{(T,T)}\right] P\text{-a.s.}$$

The proof can be deduced in the same way as the proof of Theorem 2.4.

Corollary 2.1 If $(\xi(z), \Im_z)$ is a generalized diffusion-type field with a drift coefficient $A(z,\xi)$, i.e.

$$\xi(z) = \int_{[0,z]} A(z',\xi)dz' + \int_{[0,z]} b(z',\xi)W(dz'), \tag{2.27}$$

then under conditions (a), (b), and (d) of Theorem 2.5 and the condition

$$P\left\{\int_{[0,T]^2}\left(\overline{b}^{-1}(z,\xi)A(z,\xi)\right)^2 < \infty\right\} = P\left\{\int_{[0,T]^2}\left(\overline{b}^{-1}(z,\xi)a(z,\xi)\right)^2 < \infty\right\} = 1,$$

we obtain $\mu_\xi \sim \mu_\eta$, and

$$\frac{d\mu_\eta}{d\mu_\xi}(\xi) = \exp\left\{-\int_{[0,T]^2}\left(\overline{b}^{-1}(z,\xi)\right)^2(A(z,\xi) - a(z,\xi))\xi(dz) + \frac{1}{2}\int_{[0,T]^2}\left(\overline{b}^{-1}(z,\xi)\right)^2(A^2(z,\xi) - a^2(z,\xi))dz\right\}$$

$$\frac{d\mu_\xi}{d\mu_\eta}(\eta) = \exp\left\{-\int_{[0,T]^2}\left(\overline{b}^{-1}(z,\eta)\right)^2 (A(z,\eta) - a(z,\eta))\eta(dz)\right.$$
$$\left. -\frac{1}{2}\int_{[0,T]^2}\left(\overline{b}^{-1}(z,\eta)\right)^2 (A^2(z,\eta) - a^2(z,\eta))dz\right\} \tag{2.28}$$

Theorem 2.11 *Assume that the condition (b) of Theorem 2.10 holds true, and*

(a) $P\left\{\int_{[0,T]^2}\left(\overline{b}^{-1}(z,\xi)\right)^2 (A^2(z,\xi) - a^2(z,\xi))dz < \infty\right\}$
$= P\left\{\int_{[0,T]^2}\left(\overline{b}^{-1}(z,\eta)\right)^2 (A^2(z,\eta) - a^2(z,\eta))dz < \infty\right\} = 1.$

(b) There exists a nonrandom constant K, such that

$$|a(z,x)| + |b(z,x)| \le K(1 + \|x\|_z),$$

where $\|x\|_z = \max\limits_{0<z'<z} |x(z')|$.

(c) For a nonrandom constant $\widetilde{K}$

$$|a(z,x) - a(z,\widetilde{x})| + |b(z,x) - b(z,\widetilde{x})| \le \widetilde{K}\,\|x - \widetilde{x}\|_z.$$

Then $\mu_\xi \sim \mu_\eta$, *and the respective Radon–Nikodym derivatives are given by formulas (2.28).*

Proof Put

$$\chi^{(n)}(z,x) = \begin{cases} 1, & \int_{[0,z]}\left(\overline{b}^{-1}(z',x)[A(z',x) - a(z',x)]\right)^2 dz' < n, \\ 0, & \int_{[0,z]}\left(\overline{b}^{-1}(z',x)[A(z',x) - a(z',x)]\right)^2 dz' \ge n, \end{cases}$$

$$A^{(n)}(z) = a(z,x) + \chi^{(n)}(z,x)[A(z,x) - a(z,x)].$$

Consider the differential equations

$$\xi^{(n)}(z) = \int_{[0,z]} A^{(n)}\left(z',\xi^{(n)}\right)dz' + \int_{[0,z]} b\left(z',\xi^{(n)}\right)W(dz'). \tag{2.29}$$

Since the conditions of Theorem 2.11 are satisfied, there exists a unique strong solution to (2.29). Moreover,

$$A^{(n)}(z) - a(z,x) + \chi^{(n)}(z,x)[A(z,x) - a(z,x)],$$

which implies

$$\int_{[0,T]^2}\left(\overline{b}^{-1}(z',x)\left[A^{(n)}\left(z',\xi^{(n)}\right)-a\left(z',\xi^{(n)}\right)\right]\right)^2 dz' < n\, P\text{-a.s.}$$

By Theorem 2.3 and Remark 2.5 we obtain

$$E\exp\left\{-\int_{[0,T]^2}\overline{b}^{-1}\left(z',x\right)\left[A^{(n)}\left(z',\xi^{(n)}\right)-a\left(z',\xi^{(n)}\right)\right]W\left(dz'\right)\right.$$
$$\left.-\frac{1}{2}\int_{[0,T]^2}\left(\overline{b}^{-1}\left(z',x\right)\left[A^{(n)}\left(z',\xi^{(n)}\right)-a\left(z',\xi^{(n)}\right)\right]\right)^2 dz'\right\}=1.$$

By Theorem 2.5, $\mu_{\xi^{(n)}} \sim \mu_\eta$, and

$$\frac{d\mu_{\xi^{(n)}}}{d\mu_\eta}(\eta)=\exp\left\{-\int_{[0,T]^2}\left(\overline{b}^{-1}(z,\eta)\right)^2\left(A^{(n)}(z,\eta)-a(z,\eta)\right)\eta(dz)\right.$$
$$\left.-\frac{1}{2}\int_{[0,T]^2}\left(\overline{b}^{-1}(z,\eta)\right)^2\left(\left[A^{(n)}(z,\eta)\right]^2-a^2(z,\eta)\right)dz\right\}.$$

Since the condition (a) is fulfilled, we obtain the assertion of the theorem in the same fashion as in the case of Theorem 2.4. □

Now we are interested in the following problem: how a diffusion-type field behaves under the change of the measure, and when the new transformed field is also of diffusion type in some other space. We reconstruct the form of field if it is known that the corresponding measure is absolutely continuous with respect to the measure generated by some diffusion-type field.

Consider two probability spaces $(\Omega, \Im, P)$ and $\left(\Omega, \Im, \widetilde{P}\right)$, where $d\widetilde{P}(\omega)=\rho(T,T)$ $dP(\omega)$, and

$$\rho(z)=1+\int_{[0,z]}\gamma(t_1,s)W(dz_1),\quad z\in[0,T]^2.$$

Theorem 2.12 *Assume that ξ is a solution to the stochastic differential equation*

$$\xi(z)=\int_{[0,z]}a(u,\xi)du+\int_{[0,z]}b(u,\xi)W(du), \tag{2.30}$$

where functionals $a(z,f)$ and $b(z,f)$, $z\in[0,T]^2$, belong to the space $C[0,T]^2$ and do not depend on the future. Moreover, assume that the conditions below are satisfied:

(a) $P\left\{\int_{[0,T]^2}b^2(u,\xi)du<\infty\right\}=1.$

(b) $P\left\{ \int_{[0,T]^2} \left(\overline{b}^{-1}(z',\xi)\left[a(z',\xi) + b(z',\xi)\gamma(z')\overline{\rho}^{-1}(t',s)\right]\right)^2 dz' < \infty \right\} = 1.$

(c) For almost all $z \in [0,T]^2$ $P\{b^2(z,\xi) > 0\} = 1$.

Then there exists a random field $\widetilde{\xi} = \left(\xi(z), \Im_z, \widetilde{P}\right)$ *on the probability space* $\left(\Omega, \Im, \widetilde{P}\right)$, *which is the solution to some another stochastic differential equation and, moreover, the measure* $\mu\widetilde{\xi}$ *corresponding to the field* $\widetilde{\xi}$ *is absolutely continuous with respect to the measure* μ_ξ, *generated by the field* ξ, *and the Radon–Nikodym derivative is given by*

$$\frac{d\mu\widetilde{\xi}}{d\mu_\eta} = E\left[\rho(T,T)/\Im^{\xi}_{(T,T)}\right].$$

Proof On the probability space $\left(\Omega, \Im, \widetilde{P}\right)$ the random field $\widetilde{W} = \left(\widetilde{W}(z), \Im_z\right)$, $z \in [0,T]^2$, given by

$$\widetilde{W}(z) = W(z) - \int_{[0,z]} \rho(t',s)\gamma(z')dz',$$

is a Wiener field, see Theorem 2.3. Therefore, under conditions (a) and (b) the random field $\widetilde{\xi}$ on the probability space $\left(\Omega, \Im, \widetilde{P}\right)$ is the Ito field with respect to $\widetilde{W}$ with diffusion coefficients

$$\alpha(z') = a(z',\xi) + b(z',\xi)\overline{\rho}^{-1}(t',s)\gamma(z')$$

and

$$\beta(z') = b(z',\xi).$$

In view of conditions (a)–(c) one can apply Theorem 2.1 by which there exist functionals $\widetilde{a}(z,f)$ and $\widetilde{b}(z,f)$ on the space $C[0,1]^2$, and a Wiener field $\hat{W}$ on $\left(\Omega, \Im, \widetilde{P}\right)$, such that

$$\widetilde{\xi}(z) = \int_{[0,z]} \widetilde{a}(u,\xi)du + \int_{[0,z]} \widetilde{b}(u,\xi)\hat{W}(du)\, P\text{-a.s.}$$

Let $\mu\widetilde{\xi}$ and μ_ξ be the measures corresponding, respectively, to the fields $\widetilde{\xi}$ and ξ. Then for any set $I = (x,y) \subset R$ we have

$$\mu\widetilde{\xi}(I) = \widetilde{P}\left\{\omega : \widetilde{\xi}(\omega) \in I\right\} = \int \chi_I(\xi)d\widetilde{P}(\omega) = \int \chi_I(\xi)\rho(1,1)dP(\omega)$$
$$= \int \chi_I(\xi)E\left[\rho(1,1)/\Im^{\xi}_{(1,1)}\right]dP(\omega).$$

The last equality can be expanded to all Borel sets from R^2. Put $\zeta(\xi) = E\left[\rho(T,T)/\Im^{\xi}_{(T,T)}\right]$. Thus, we proved

$$\mu\widetilde{\xi}(I) = \int_I \zeta(f)d\mu(f),$$

which implies the assertion of the theorem. □

Assume now that the random field ξ is a solution to (2.30), and the field ζ is such that $\mu_\zeta \ll \mu_\xi$ with

$$\frac{d\mu_\zeta}{d\mu_\xi}(z,\xi) = \rho_1(z,\xi) = 1 + \int_{[0,z]} \beta(u)W(du).$$

Put $\eta(z) = \int_{[0,z]} b(u,\xi)W(du)$. Theorem 2.3 asserts that under conditions

B1. for any $z \in [0,T]^2$ the equation $b(z,\xi)\alpha(z) = a(z,\xi)$ has (with respect to α) P-a.s. a bounded solution and

$$E\exp\left\{-\int_{[0,T]^2} \alpha(z)W(dz) - \frac{1}{2}\int_{[0,T]^2} \alpha^2(z)dz\right\} = 1,$$

B2. $P\left\{\int_{[0,T]^2} \left(\overline{b}^{-1}(u,\xi)a(u,\xi)\right)^2 du < \infty\right\} = 1$

the measures μ_ξ and μ_η are equivalent, and the Radon–Nikodym derivative is given by

$$\frac{d\mu_\xi}{d\mu_\eta}(z,\eta) = \rho_2(z,\eta)$$

$$= \exp\left\{\int_{[0,z]} \overline{b}^{-1}(u,\eta)a(u,\eta)W(du) - \int_{[0,z]} \left(\overline{b}^{-1}(u,\eta)a(u,\eta)\right)^2 du\right\}.$$

Appling the Ito formula (see Theorem 1.17) we derive

$$\rho_2(z,\eta) = 1 + \int_{[0,z]} \overline{b}^{-1}(z',\eta)a(z',\eta)\rho_2(t',s,\eta)W(dz'),$$

implying $\mu_\zeta \ll \mu_\xi \ll \mu_\eta$. Let us calculate the derivative $\rho_3(z,\eta) = \frac{d\mu_\zeta}{d\mu_\xi}$. We have:

$$\begin{aligned}\rho_3(z,\eta) = 1 &+ \int_{[0,z]} \left(\beta(z')\rho_2(z',\eta) + \rho_1(z',\eta)\rho_2(t',s,\eta)\overline{b}^{-1}(z',\eta)a(z',\eta)\right)W(dz') \\ &+ \int_{[0,z]}\int_{[0,z]} \left(\beta(x,s')\rho_2\left(t,'s,\eta\right)\overline{b}^{-1}\left(t,'y,\eta\right)a(z',\eta)\right)W(dx,dy)W(dz') \\ &+ \int_{[0,z]}\int_{[0,z']} \left(\beta(t',y)\rho_2(x,s,\eta)\overline{b}^{-1}(x,s',\eta)a(z',\eta)\right)W(dx,dy)W(dz').\end{aligned}$$

Suppose that the random fields $a,\ b,\ \beta$ are such that

$$\int_{[0,z]}\int_{[0,z]}\left(\beta(x,s')\rho_2\left(t,'s,\eta\right)\overline{b}^{-1}\left(t,'y,\eta\right)a(z',\eta)\right)W(dx,dy)W(dz')$$
$$+\int_{[0,z]}\int_{[0,z']}\left(\beta(t',y)\rho_2(x,s,\eta)\overline{b}^{-1}(x,s',\eta)a(z',\eta)\right)W(dx,dy)W(dz')=0.$$

Then applying again the Ito formula we derive

$$\rho_3(z,\eta)=1+\int_{[0,z]}\alpha(u)W(du),$$

$$\alpha(z')=\beta(z')\rho_2(z',\eta)+\rho_1(z',\eta)\rho_2(t',s,\eta)\overline{b}^{-1}(z',\eta)a(z',\eta).$$

Thus, $\rho_3(z,\eta)$ is nonnegative martingale with $E\rho(T,T)=1$. For an $\mathfrak{I}^{\eta}_{z'}$-measurable random variable $\lambda=\lambda(\eta)$ we have

$$E\lambda(\eta)\rho(z,\eta)=\int\lambda(f)d\mu_\zeta(z,f)=\int\lambda(f)\rho_3(z',f)d\mu_\eta(z,f).$$

Consider the probability space $\left(\Omega,\mathfrak{I}^{\eta},\widetilde{P}\right)$, with $d\widetilde{P}(\omega)=\rho(T,T)dP(\omega)$. The random field

$$\widetilde{W}(z)=W(z)-\int_{[0,z]}\overline{\rho}_3^{\,-1}(t',s,\eta)b(z',\eta)\alpha(z')dz'$$

is a Wiener field on this space, hence by Theorem 2.1 there exists a functional $\widetilde{a}(z,f)$ in the space $C[0,T]^2$ such that for almost all $z'\in[0,T]^2$

$$\widetilde{a}(z',f)=\overline{\rho}_3^{\,-1}(t',s,\eta)b(z',\eta)\alpha(z')$$

and

$$\widetilde{P}\left\{\int_{[0,T]^2}\widetilde{a}^2(u,\eta)du<\infty\right\}=1.$$

Moreover,

$$P\left\{\int_{[0,T]^2}\widetilde{a}^2(\mu,\eta)du<\infty\right\}=\mu_\zeta\left\{\int_{[0,T]^2}\widetilde{a}^2(\mu,f)du<\infty\right\}$$
$$=\widetilde{P}\left\{\int_{[0,T]^2}\widetilde{a}^2(u,\eta)du<\infty\right\}=1,$$

and in such a way all stochastic integrals written above are well defined.

Define on the probability space $\left(\Omega, \Im^{\eta}, \widetilde{P}\right)$ a random field $\left(\varphi(z,\zeta), \Im_z^{\zeta}\right)$, $z \in [0,T]^2$, by

$$\varphi(z,f) = f(z) - \int_{[0,z]} \widetilde{a}^2(u,f)du, \quad f \in C[0,T]^2.$$

Let $\lambda(\zeta)$ be a bounded random $\Im_z^{\zeta*}$-measurable variable. Then for any $z_1 \in [0,T]^2$, $z_1 > z$, we have

$$\begin{aligned} E\lambda(\zeta)\,\varphi(z,z_1](\zeta) &= \int \lambda(f)\,\phi(z,z_1](f)d\mu_{\zeta}(f) = \int \lambda(f)\,\phi(z,z_1](f)\,\rho(T,,T)d\mu_{\eta}(f) \\ &= \int \lambda(\eta)\,\phi(z,z_1](\eta)\,\rho(T,T)dP = \widetilde{E}\,\lambda(\eta)\widetilde{E}\left[\phi(z,z_1](\eta)/\Im_z^{\zeta*}\right] = 0. \end{aligned}$$

Therefore, $\zeta(z) - \int_{[0,z]} \widetilde{a}^2(u,\zeta)du$ is a strong martingale, and there exists (see Theorems 1.12, 2.1) $\phi(z,f)$ such that for almost all $f \in C[0,1]^2$

$$\int_{[0,T]^2} \varphi^2(u,f)du < \infty$$

and

$$\zeta(z) = \int_{[0,z]} \widetilde{a}(u,\zeta)du - \int_{[0,z]} \phi(u,\zeta)\widetilde{W}(du), \tag{2.31}$$

where $\widetilde{W}$ is a Wiener field on $\left(\Omega, \Im\widetilde{P}\right)$. In such a way, we proved.

Theorem 2.13 *Assume that ξ is a strong solution to (2.30) with coefficients $a(z,f)$ and $b(z,f)$, satisfying B1 and B2. Assume that the stochastic field ζ is such that $\mu_\zeta \ll \mu_\xi$, and the Radon–Nikodym derivative is given by*

$$\frac{d\mu_\zeta}{d\mu_\xi}(z,\xi) = 1 + \int_{[0,z]} \beta(t_1,s)W(dz_1).$$

In addition, assume that $a(z,f)$, $b(z,f)$ and $\beta(z)$ satisfy

$$\begin{aligned} &\int_{[0,z]}\int_{[0,z]} \left(\beta(x,s')\rho_2\left(t,'s,\eta\right)\overline{b}^{-1}\left(t,'y,\eta\right)a(z',\eta)\right)W(dx,dy)W(dz') \\ &+ \int_{[0,z]}\int_{[0,z']} \left(\beta(t',y)\rho_2(x,s,\eta)\overline{b}^{-1}(x,s',\eta)a(z',\eta)\right)W(dx,dy)W(dz') = 0 \end{aligned}$$

Then there exist a probability space $\left(\Omega, \Im, \widetilde{P}\right)$, the coefficients $\widetilde{a}(z,f)$ and $\widetilde{b}(z,f)$, the field $\phi(z,f)$ and a Wiener field $\widetilde{W}$, such that random field ζ is the solution to the stochastic differential equation (2.31).

2.5 Nonparametric Estimation of a Two-Parametrical Signal from Observation with Additive Noise

In this section we study the properties of a periodic in both variables estimator, observed on enlarged part of the plane, with random errors of white noise type. Here we present results obtained in [13, 15, 17, 42–44].

Suppose that on the probability space $(\Omega, \Im, P)$ we have a real random field $(x(z),\ z \in R_+^2)$ and continuous square integrable strong martingale $(\xi(z),\ z \in R_+^2)$. We consider the problem of estimating the unknown function a from observations of the two-dimensional random field $(x(z), y(z),\ z \in R_+^2)$ in the rectangle $[0,T] \times [0,S]$ where

$$y(z) = \int_{[0,z]} a_0(u)x(u)du + \xi(z). \tag{2.32}$$

Further we assume that the random fields $x(z)$ and $\xi(z)$ are independent and the conditions below are satisfied:

C1. The function a_0 is an element of the set K of all real functions defined on the plane, 2π-periodic in both variables, whose Fourier coefficients

$$c_{kl}(a) = \frac{1}{4\pi^2}\int_0^{2\pi}\int_0^{2\pi} a(t,s)\exp\{i(kt+ls)\}dtds, \quad k,l = 0, \pm 1, \pm 2, \ldots$$

satisfy the inequalities

$$|c_{00}(a)| \le L,\ |c_{k0}(a)||k|^{\alpha} \le L,\ |c_{0l}(a)||l|^{\beta} \le L,\ |c_{kl}(a)||k|^{\alpha}|l|^{\beta} \le L,\ kl \ne 0,$$

with some constants $L > 0$, $\alpha > 3$, $\beta > 3$.

For any function $a \in K$ put $\|a\|^2 = \frac{1}{4\pi^2}\int_0^{2\pi}\int_0^{2\pi} a(t,s)dtds$. We call the element $a \in K$ an interior point of K, if $|c_{00}(a)| \le \widetilde{L}$, $|c_{k0}(a)||k|^{\alpha} \le \widetilde{L}$, $|c_{0l}(a)|$ $|l|^{\beta} \le \widetilde{L}$, $|c_{kl}(a)||k|^{\alpha}|l|^{\beta} \le \widetilde{L}$, $kl \ne 0$, with some $\widetilde{L} < L$.

C2. $(\xi(z),\ z \in R_+^2)$, $\xi(t,0) = \xi(0,s) = 0$ is a continuous strong martingale, square integrable in any finite rectangle, with characteristic $\gamma(z) = \int_{[0,z]}\sigma^2(u)du$, where $\sigma^2(z) > 0$ P- almost surely and $E\sigma^2(z) \le C$.

C3. $(x(z),\ z \in R_+^2)$ is a real random field whose sample functions have continuous second-order derivatives with P-probability 1. The random field $(x^2(z),\ z \in R_+^2)$ is homogeneous in the broad sense, with $Ex^2(0) > 0$.

Denote by $r(z) = E([x^2(z) - Ex^2(0)][x^2(0) - Ex^2(0)])$ the correlation function of the field $(x^2(z),\ z \in R_+^2)$.

C4. For some $\gamma_1 > 0$, $\gamma_2 > 0$, $L_1 > 0$ and all $S \ge 1$, $T \ge 1$ the inequality $\int_0^T\int_0^S |r(z)|dz \le L_1 T^{1-\gamma_1} S^{1-\gamma_2}$ holds true.

C5. For some positive $L_2 > 0$ $|r(z)| \le \dfrac{L_2}{(1+t^2)(1+s^2)}$, $z = (t,s) \in R_+^2$.

The conditions C1–C3 guarantee that for any function from the set K one can define a stochastic integral with respect to a strong martingale.

We consider the problem of estimation a_0 from the observations ($x(z)$, $y(z)$, $z \in [0,T] \times [0,S]$) of the stochastic field ($y(z)$, $z \in R_+^2$), defined by (2.32) with some fixed function $a \in K$. As an estimate for a_0 we take an element $a_{TS} \in K$ which minimizes the functional

$$Q_{TS} = \frac{1}{TS}\int_0^T\int_0^S a(u)x(u)y(du) - \frac{1}{2TS}\int_0^T\int_0^S a^2(u)x^2(u)du$$

on K. In assumption C1 the maximum of Q_{TS} on K is achieved, and one can show using Theorem 1.20 that $\{a_{TS}(z), z \in R_+^2\}$ is a separable measurable field. The functional Q_{TS} can be represented as

$$Q_{TS} = \frac{1}{TS}\int_0^T\int_0^S [a(u) - a_0(u)]x(u)\xi(du) - \frac{1}{2TS}\int_0^T\int_0^S [a(u) - a_0(u)]^2 x^2(u)du + Q_{TS}(a_0).$$

Further, to simplify our calculation we put $T = S$ and denote $a_{TT} := a_T$, $Q_{TT} := Q_T$.

Lemma 2.9 *Assume that conditions C1–C3 hold true. Then for any $\gamma > 0$*

$$P\left\{\lim_{T\to\infty}\max_{a\in K}\left|\frac{1}{T^{1+\gamma}}\int_{[0,T]^2} a(u)x(u)\xi(du)\right| = 0\right\} = 1.$$

Proof For $T > 0$ denote

$$\eta_T = \max_{a\in K}\left|\frac{1}{T^{1+\gamma}}\int_{[0,T]^2} a(u)x(u)\xi(du)\right|.$$

Let us estimate $E\eta_T{}^2$. From the Fourier decomposition of the function a and condition C1 we obtain

$$E\eta_T^2 \le E\left\{\max_{a\in K}\left|\sum_{j,k=-\infty}^{\infty} c_{jk}(a)\frac{1}{T^{1+\gamma}}\int_{[0,T]^2}\exp\{i(jt' + ks')\}x(z')\xi(dz')\right|^2\right\}$$
$$\le E\left\{\left|\sum_{j,k=-\infty}^{\infty}{}' c_{jk}(a)\frac{1}{T^{1+\gamma}}\int_{[0,T]^2}\exp\{i(jt' + ks')\}x(z')\xi(dz')\right|^2\right\}.$$

Here the dash near sum sign means that we take 1 instead of $|j|^{-\alpha}$ as $j = 0$ and $|k|^{-\beta}$ as $k = 0$. By the properties of the stochastic integrals we get

$$
\begin{aligned}
E\eta_T^2 &\leq \left\{ \sum_{j,k=-\infty}^{\infty} \frac{L}{T^{1+\gamma}|j|^{\alpha}|k|^{\beta}} \left[E\left| \int_{[0,T]^2} \exp\{i(jt'+ks')\}x(z')\xi(dz') \right|^2 \right]^{1/2} \right\}^2 \\
&\leq C^* T^{-2\gamma},
\end{aligned}
\tag{2.33}
$$

where

$$
C^* = \left(\sum_{j,k=-\infty}^{\infty} \frac{L}{|j|^{\alpha}|k|^{\beta}} \right)^2 Ex^2(0)C.
$$

Let p be a fixed natural number such that $2p\gamma > 1$. From the previous estimate we obtain using the Borel-Cantelli lemma for the sequence $\{\eta_{T(n)},\ n \geq 1\}$, with $T(n) = n^p, p \geq 1$, that $P\left\{\lim_{n\to\infty} \eta_{T(n)} = 0\right\} = 1$. Take now $T \in [T(n), T(n+1)]$. Then $\eta_T = \eta_{T(n)} + \zeta_n$, where $\zeta_n = \zeta_{1n} + \zeta_{2n}$, and

$$
\zeta_{1n} = \frac{1}{T^{1+\gamma}(n)} \max_{T\in[T(n),\,T(n+1)]} \max_{a\in K} \left| \int_{T(n)}^{T} \int_0^T a(u)x(u)\xi(du) \right|,
$$

$$
\zeta_{2n} = \frac{1}{T^{1+\gamma}(n)} \max_{T\in[T(n),\,T(n+1)]} \max_{a\in K} \left| \int_0^{T(n)} \int_{T(n)}^{T} a(u)x(u)\xi(du) \right|.
$$

Let us estimate $E\zeta_{1n}$. We have

$$
\begin{aligned}
E\zeta_{1n} &\leq \frac{1}{T^{2(1+\gamma)}(n)} E\left[\sum_{j,k=-\infty}^{\infty} \frac{L}{|j|^{\alpha}|k|^{\beta}} \max_{T\in[T(n),T(n+1)]} \left| \int_{T(n)}^{T} \int_0^T \exp\{i(jt'+ks')\}x(z')\xi(dz') \right| \right]^2 \\
&\leq \frac{1}{T^{2(1+\gamma)}(n)} \left[\sum_{j,k=-\infty}^{\infty}{}' \frac{L}{|j|^{\alpha}|k|^{\beta}} \left\{ E \max_{T\in[T(n),T(n+1)]} \left| \int_{T(n)}^{T} \int_0^T \exp\{i(jt'+ks')\}x(z')\xi(dz') \right|^2 \right\}^{1/2} \right]^2.
\end{aligned}
$$

It is easy to see from (2.33) that

$$
\begin{aligned}
&E\left[\max_{T\in[T(n),\,T(n+1)]} \left| \int_{T(n)}^{T} \int_0^T \exp\{i(jt'+ks')\}x(z')\xi(dz') \right|^2 \right] \\
&\quad \leq 16T(n+1)[T(n+1)-T(n)]Ex^2(0)c.
\end{aligned}
$$

Thus,

$$E\zeta_{1n} \leq 16c_1 \frac{T(n+1)[T(n+1)-T(n)]}{T^{2(1+\gamma)}(n)} = 16c_1 \frac{T(n+1)-T(n)}{T^{1+2\gamma}(n)} \frac{T(n+1)}{T(n)}$$
$$= \frac{16c_1}{n^{2p\gamma}} \left[\left(1+\frac{1}{n}\right)^p - 1 \right] \left(\frac{n+1}{n}\right)^p.$$

Hence, $P\{\lim_{n\to\infty} \zeta_{1n} = 0\} = 1$. Similarly, one can show that $P\{\lim_{n\to\infty} \zeta_{2n} = 0\} = 1$. Lemma is proved. □

Remark 2.6 The assertions of Lemma 2.8 also hold true also if instead of the function $a \in K$ a difference of two functions from K is considered.

Lemma 2.10 *Assume that* $(\varsigma(z),\, z \in R^2)$ *is a real homogeneous field with zero mean and correlation function* $r(z) = E(\varsigma(z)\varsigma(0))$, $z \in R^2$, *such that for all* $T \geq 1$ *and some positive* L_3 *and* δ

$$\int_{[0,T]^2} |r(z)| dz \leq L_3 T^{2-\delta}.$$

Then

$$P\left\{ \lim_{T\to\infty} \max_{a\in K} \left| \frac{1}{T^2} \int_{[0,T]^2} a(u)\varsigma(u)du \right| = 0 \right\} = 1.$$

Proof For any $T > 0$ let

$$\widetilde{\eta}_T = \sup_{a\in K} \left| \frac{1}{T^2} \int_{[0,T]^2} a(u)\varsigma(u)du \right|,$$

and for $n \geq 1$ let $\widetilde{\zeta}_n = \max_{T\in[T(n),\, T(n+1)]} \widetilde{\eta}_T$ where $T(n) = n^p, p \geq 1, p$ is the fixed value with $\delta p > 1$. Decompose a into the Fourier series:

$$\widetilde{\eta}_T = \sup_{a\in K} \left| \sum_{j,k=-\infty}^{\infty} c_{jk}(a) \frac{1}{T^2} \int_{[0,T]^2} \exp\{i(jt' + ks')\}\varsigma(z')dz' \right|$$
$$\leq \sum_{j,k=-\infty}^{\infty} \frac{L}{|j|^\alpha |k|^\beta} \left| \frac{1}{T^2} \int_{[0,T]^2} \exp\{i(jt' + ks')\}\varsigma(z')dz' \right|.$$

The series in the right-hand side of the last formula converge with P-probability 1. For $\widetilde{\zeta}_n$ we have

$$\widetilde{\zeta}_n \leq \sum_{j,k=-\infty}^{\infty}{}' \frac{L}{T^2(n)\,|j|^\alpha |k|^\beta} \left| \int_{[0,T(n)]^2} \exp\{i(jt' + ks')\}\varsigma(z')dz' \right|$$
$$+ \sum_{j,k=-\infty}^{\infty}{}' \frac{L}{T^2(n)\,|j|^\alpha |k|^\beta} \int_{T(n)}^{T(n+1)} \int_0^{T(n+1)} |\varsigma(z')|dz'$$
$$+ \sum_{j,k=-\infty}^{\infty}{}' \frac{L}{T^2(n)\,|j|^\alpha |k|^\beta} \int_0^{T(n)} \int_{T(n)}^{T(n+1)} |\varsigma(z')|dz' = \widetilde{\zeta}_{1n} + \widetilde{\zeta}_{2n} + \widetilde{\zeta}_{3n}.$$

Let us estimate $E\left(\widetilde{\zeta}_{1n}\right)^2$. Observe that

$$E\left(\widetilde{\zeta}_{1n}\right)^2 \leq \left\{ \sum_{j,k=-\infty}^{\infty} \frac{L}{|j|^\alpha |k|^\beta} \left[E\left| \frac{1}{T^2(n)} \int_{[0,T(n)]^2} \exp\{i(jt' + ks')\}\varsigma(z')dz' \right|^2 \right]^{1/2} \right\}^2.$$

By the conditions of Lemma 2.9 we have

$$E\left\{ \frac{1}{T^4(n)} \left| \int_{[0,T(n)]^2} \exp\{i(jt' + ks')\}\varsigma(z')dz' \right|^2 \right\}$$
$$\leq \frac{1}{T^4(n)} \int_0^{T(n)} \int_0^{T(n)} \int_0^{T(n)} \int_0^{T(n)} |r(t-t', s-s')|dzdz'$$
$$\leq \frac{1}{T^2(n)} \int_{[0,T]^2} |r(z)|dz \leq \frac{4L_3}{T^\delta},$$

which implies

$$E\left(\widetilde{\zeta}_{1n}\right)^2 \leq \frac{c_2}{T^\delta(n)} = \frac{c_2}{n^{\delta p}},$$

where $c_2 = 4L_3 \left(\sum_{j,k=-\infty}^{\infty} \frac{L}{|j|^\alpha |k|^\beta} \right)^2$.

Then from the Borel-Cantelli lemma we deduce that $P\left\{\lim_{n\to\infty} \widetilde{\zeta}_{1n} = 0\right\} = 1$. Similarly, one can show that $P\left\{\lim_{n\to\infty} \widetilde{\zeta}_{2n} = 0\right\} = 1$ and $P\left\{\lim_{n\to\infty} \widetilde{\zeta}_{3n} = 0\right\}$ $= 1$, which proves the assertion of the lemma. □

Remark 2.7 Lemmas 2.8 and 2.9 also hold true if the function $a \in K$ is substituted with $a(z)b(z)$, where b is a bounded deterministic function, or a random function with $E|b(z)|^2 \leq L$, where L is some constant independent of $\xi(z)$.

Theorem 2.14 *Assume that conditions C1–C4 hold true. Then*

$$P\left\{\lim_{T\to\infty}\max_{z\in R^2}|a_T(z)-a_0(z)|=0\right\}=1.$$

Proof By the definition of a_T we have $Q_T(a_T)=\max_{a\in K} Q_T(a)$. Therefore, $Q_T(a_T)\geq Q_T(a_0)$, and thus

$$\begin{aligned}Q_T(a_T)-Q_T(a_0)&=\frac{1}{T^2}\int_{[0,T]^2}[a_T(z)-a_0(z)]x(z)\xi(dz)\\&-\frac{1}{2T^2}\int_{[0,T]^2}[a_T(z)-a_0(z)]^2x^2(z)dz\geq 0.\end{aligned}$$

The last inequality implies that

$$\begin{aligned}&\max_{a\in K}\left|\frac{1}{T^2}\int_{[0,T]^2}[a_T(z)-a_0(z)]x(z)\xi(dz)\right|\\&+\max_{a\in K}\left|\frac{1}{2T^2}\int_{[0,T]^2}[a_T(z)-a_0(z)]^2\left[x^2(z)-Ex^2(0)\right]dz\right|\\&\geq\frac{1}{2T^2}\int_{[0,T]^2}[a_T(z)-a_0(z)]^2dz\cdot Ex^2(0).\end{aligned}$$

Both summands in the left-hand side of the last inequality converge to zero as $T\to\infty$ P-almost surely, which follows, respectively, from Lemma 2.8 and Lemma 2.9. Hence,

$$P\left\{\lim_{T\to\infty}\frac{1}{T^2}\int_{[0,T]^2}[a_T(z)-a_0(z)]^2dz=0\right\}=1,$$

and in such a way

$$P\left\{\lim_{T\to\infty}\|a_T(z)-a_0(z)\|=0\right\}=1.$$

Taking into account the Hölder inequality and the fact that the set K is compact with respect to the uniform convergence, we obtain

$$P\left\{\lim_{T\to\infty}\max_{z\in R^2}|a_T(z)-a_0(z)|=0\right\}=1.\qquad\square$$

Consider now the asymptotic distribution of some functionals that depend on $a_T(z)$ and convergence of measures generated by these estimators. Further we assume that the square integrable functional under consideration is Gaussian.

Theorem 2.15 *Let* b *be a real function defined on the plane, 2π-periodical with respect to both parameters. Assume that* b *satisfies the conditions below:*

(a) b is continuous (may be with exception of finite number of interval $t =$ const or $s =$ const) and bounded on $[0, 2\pi]^2$

(b) $\|b\| = 1$

(c) $\int_{[0,2\pi]^2} b(z)a_0(z)dz$ is the inner point of the set $I = \left\{ \int_{[0,2\pi]^2} b(z)a(z)dz, \quad a \in K \right\}$

Assume also that

$$\lim_{T\to\infty} \frac{1}{T^2} \int_{[0,2\pi]^2} b^2(z)E\sigma^2(z)\,dz = \beta,$$

where σ^2 is defined in condition C2, and that conditions C1–C5 are fulfilled.

Then the distribution of the random variable $T\int_{[0,2\pi]^2} [a_T(z) - a_0(z)]b(z)dz$ converges weakly as $T \to \infty$ to the Gaussian distribution with zero mean and variance $\beta(Ex^2(0))^{-1}$.

Proof For any point $a \in K$ we define a real variable $\theta(a)$ as

$$\theta(a) = \frac{1}{4\pi^2} \int_{[0,2\pi]^2} a(z)b(z)dz.$$

The function a can be written as

$$a(z) = \theta(a)b(z) + q_a(z), \quad z \in R^2,$$

where $q_a = a - \theta(a)b$, and $\int_{[0,2\pi]^2} q_a(z)b(z)dz = 0$.

Consider the functional $Q_T(a)$, $a \in K$, as a function of the real parameter $\theta(a)$, $a \in K$:

$$\begin{aligned} Q_T(\theta(a)) = Q_T(a) &= \frac{1}{T^2} \int_{[0,T]^2} [\theta(a)b(z) + q_a(z)]x(z)\xi(dz) \\ &+ \frac{1}{T^2} \int_{[0,T]^2} [\theta(a)b(z) + q_a(z)]a_0(z)x^2(z)dz \\ &- \frac{1}{2T^2} \int_{[0,T]^2} [\theta(a)b(z) + q_a(z)]x^2(z)dz. \end{aligned}$$

Since a_T $Q_T(a_T) \geq Q_T(a)$, $a \in K$, we have

$$Q_T(\theta(a_T)) \geq Q_T(\theta(a_0)). \qquad (2.34)$$

Under the conditions of lemma, the conditions of Theorem 2.14 are fulfilled, implying $P\left\{\lim_{T\to\infty}\theta(a_T)=\theta(a_0)\right\}=1$. From condition (c) we obtain that the variable $\theta(a_T)$ is the inner point of the region I with probability tending to 1 as $T\to\infty$. Moreover, we obtain from (2.34) that with the same probability $\theta(a_T)$ satisfies

$$\frac{1}{T^2}\int_{[0,T]^2} b(z)x(z)\,\xi(dz)=[\theta(a_T)-\theta(a_0)]\frac{1}{T^2}\int_{[0,T]^2} b^2(z)x^2(z)dz$$
$$+\frac{1}{T^2}\int\limits_{[0,T]^2} b(z)\left[q_{a_T}(z)-q_{a_0}(z)\right]x^2(z)dz.$$

Therefore the limit distribution of the variable $T[\theta(a_T)-\theta(a_0)]$ coincides (as $T\to\infty$) with the limit distribution of $T\left[\widetilde{\theta}\,(a_T)-\widetilde{\theta}\,(a_0)\right]$, given by

$$T\left[\widetilde{\theta}\,(a_T)-\widetilde{\theta}\,(a_0)\right]=\left[\frac{1}{T^2}\int_{[0,T]^2} b^2(z)x^2(z)dz\right]^{-1}$$
$$\times\left(\frac{1}{T}\int_{[0,T]^2} b(z)x(z)\,\xi(dz)-\frac{1}{T}\int_{[0,T]^2} b(z)\left[q_{a_T}(z)-q_{a_0}(z)\right]x^2(z)dz\right). \tag{2.35}$$

Note that under conditions of Lemma 2.10 all conditions of Lemma 2.9 are also fulfilled, implying that $\frac{1}{T^2}\int_{[0,T]^2} b^2(z)x^2(z)dz\to Ex^2(0)$ in probability as $T\to\infty$. Thus, it is enough to show that the integral

$$\frac{1}{T}\int_{[0,T]^2} b(z)\left[q_{a_T}(z)-q_{a_0}(z)\right]x^2(z)dz=\eta_{1T}+\eta_{2T}+\eta_{3T}+\eta_{4T}$$

converges to 0 in probability as $T\to\infty$ and $N=[T/2\pi]$. Put

$$\eta_{1T}=\frac{1}{T}\int_{2\pi N}^{T}\int_{2\pi N}^{T} b(z)\left[q_{a_T}(z)-q_{a_0}(z)\right]x^2(z)dz,$$
$$\eta_{2T}=\frac{1}{T}\int_{0}^{2\pi}\int_{2\pi N}^{T} b(z)\left[q_{a_T}(z)-q_{a_0}(z)\right]x^2(z)dz,$$
$$\eta_{3T}=\frac{1}{T}\int_{2\pi N}^{T}\int_{0}^{2\pi N} b(z)\left[q_{a_T}(z)-q_{a_0}(z)\right]x^2(z)dz,$$
$$\eta_{4T}=\frac{1}{T}\int_{0}^{2\pi N}\int_{0}^{2\pi N} b(z)\left[q_{a_T}(z)-q_{a_0}(z)\right]x^2(z)dz.$$

It is easy to show directly that η_{1T} converges to 0 in probability as $T \to \infty$. For η_{2T} and η_{3T} the idea of the proof is the same. Decompose

$$\eta_{2T} = \frac{1}{T}\int_0^{2\pi}\int_{2\pi N}^{T} b(z)\left[q_{a_T}(z) - q_{a_0}(z)\right]x^2(z)dz$$
$$= \frac{1}{T}\int_0^{2\pi}\int_{2\pi N}^{T} b(z)\left[q_{a_T}(z) - q_{a_0}(z)\right]\left[x^2(z) - Ex^2(0)\right]dz$$
$$+\frac{1}{T}\int_0^{2\pi}\int_{2\pi N}^{T} b(z)\left[q_{a_T}(z) - q_{a_0}(z)\right]Ex^2(0)dz = \eta'_{2T} + \eta''_{2T}.$$

From the Cauchy inequality we deduce, performing the necessary transformations, that

$$E\left|\eta''_{2T}\right| \le Ex^2(0)\frac{1}{T}\left[\int_0^{2\pi N}\int_0^{2\pi N} b^2(z)dz\right]^{1/2}$$
$$\times\left(\int_0^{2\pi N}\int_0^{2\pi} E\left[q_{a_T}(z) - q_{a_0}(z)\right]^2 dz\right)^{1/2}$$
$$\le \sqrt{2\pi}Ex^2(0)\left(\frac{1}{2\pi}\int_0^{2\pi}\int_0^{2\pi} E\left[q_{a_T}(z) - q_{a_0}(z)\right]^2 dz\right)^{1/2}.$$

For any fixed z we have by the Lebesgue theorem and the fact that $q_{a_T}(z) - q_{a_0}(z) \to 0$ as $T \to \infty$ P-a.s

$$E\left[q_{a_T}(z) - q_{a_0}(z)\right]^2 \to 0 \text{ as } T \to \infty.$$

Therefore, by the Lebesgue theorem we have $E|\eta''_{2T}| \to 0$ as $T \to \infty$. For $E|\eta'_{2T}|$ we obtain

$$E\left|\frac{1}{T}\int_0^{2\pi}\int_{2\pi N}^{T} b(z)\left[q_{a_T}(z) - q_{a_0}(z)\right]\left[x^2(z) - Ex^2(0)\right]dz\right|$$
$$\le E\left|\frac{1}{N}\int_0^{2\pi}\int_{2\pi N}^{T} b(z)\left[q_{a_T}(z) - q_{a_0}(z)\right]\sum_{k=0}^{N-1}\left[x^2(t+2k\pi, s) - Ex^2(0)\right]dz\right|$$
$$\le \left(\int_0^{2\pi}\int_{2\pi N}^{T} b(z)\left[q_{a_T}(z) - q_{a_0}(z)\right]^2 dz\right)^{1/2}$$
$$\times\left(\int_0^{2\pi}\int_{2\pi N}^{T} E\left[\frac{1}{N}\sum_{k=0}^{N-1}\left[x^2(t+2k\pi, s) - Ex^2(0)\right]\right]dz\right)^{1/2}.$$

Similarly to the case of $E|\eta''_{2T}|$, it is easy to show that the first multiplier in the right-hand side part of the last inequality converges to zero as $T \to \infty$. The second multiplier is uniformly bounded by $T \geq 2\pi$. Therefore, $E|\eta'_{2T}| \to 0$ as $T \to \infty$, which implies that $\eta'_{2T} \to 0$ in probability as $T \to \infty$.

The equality $\int_0^{2\pi N}\int_0^{2\pi N} b(z)\left[q_{a_T}(z) - q_{a_0}(z)\right]dz = 0$ allows to represent η_{4T} as

$$\eta_{4T} = \frac{1}{T}\int_0^{2\pi N}\int_0^{2\pi N} b(z)\left[q_{a_T}(z) - q_{a_0}(z)\right]\left[x^2(z) - Ex^2(0)\right]dz$$
$$= \int_0^{2\pi N}\int_0^{2\pi N} b(z)\left[q_{a_T}(z) - q_{a_0}(z)\right]\frac{1}{T}\sum_{j,k=0}^{N-1}\left[x^2(t+2j\pi, s+2k\pi) - Ex^2(0)\right]dz.$$

From this expression we obtain by the Cauchy inequality

$$E|\eta_{4T}| \leq \left(\int_0^{2\pi N}\int_0^{2\pi N} b^2(z)\left[q_{a_T}(z) - q_{a_0}(z)\right]^2 dz\right)^{1/2}$$
$$\times\left[\int_0^{2\pi N}\int_0^{2\pi N} E\left(\frac{1}{N}\sum_{j,k=0}^{N-1}\left[x^2(t+2j\pi, s+2k\pi) - Ex^2(0)\right]\right)^2 dz\right]^{1/2}.$$

The first multiplier in the right-hand side of the last inequality converges to zero as $T \to \infty$. For the second multiplier we obtain using condition B5

$$E\left(\frac{1}{N}\sum_{j,k=0}^{N-1}\left[x^2(t+2j\pi, s+2k\pi) - Ex^2(0)\right]\right)^2$$
$$= \frac{1}{N^2}\sum_{j_1,j_2=0}^{N-1}\sum_{k_1,k_2=0}^{N-1}$$
$$\times E\left[(x^2(t+2j_1\pi, s+2k_1\pi) - Ex^2(0))(x^2(t+2j_2\pi, s+2k_2\pi) - Ex^2(0))\right]$$
$$\leq \frac{L_2}{N^2}\sum_{j_1,j_2=0}^{N-1}\sum_{k_1,k_2=0}^{N-1}\frac{1}{1+(j_1-j_2)^2 4\pi^2}\frac{1}{1+(k_1-k_2)^2 4\pi^2} \leq c,$$

where c is a constant, independent of N. Thus, $E|\eta_{4T}| \to 0$ as $T \to \infty$, which in turn implies that $\eta_{4T} \to 0$ in probability as $T \to \infty$.

To show that $\lambda_T = \frac{1}{T}\int_{[0,T]^2} b(z)x(z)\,\xi(dz)$ is asymptotically normal as $T \to \infty$, we consider its characteristic function

$$\varphi_T(u) = E\exp\{iu\lambda_T\} = E\{E_x[\exp(iu\lambda_T)]\},$$

where E_x is the conditional expectation with respect to $\sigma\{x(z), z \in R_+^2\}$. Since $\xi(z)$ is Gaussian, its characteristic $\gamma(z)$ can be represented in the form $\gamma(z) = \int_{[0,z]}\sigma^2(z_1)dz_1$. Therefore,

$$\varphi_T(u) = E\exp\left\{-\frac{u^2}{2T^2}\int_{[0,T]^2} b^2(z)x^2(z)E\sigma^2(z)dz\right\}, \quad u\in R, P\text{-a.s.}$$

Using condition (c), Lemma 2.9, and applying the Lebesgue theorem, we obtain for any $u \in R$

$$\lim_{T\to\infty} \varphi_T(u) = \exp\left\{-\frac{u^2}{2}\beta Ex^2(0)\right\}.$$

Thus, the variable λ_T is asymptotically normal, and since η_{iT}, $i = 1, 2, 3, 4$ converge to zero as $T \to \infty$, we deduce the statement of the Theorem from (2.35) and the expression above. □

2.6 Identification Problem for Stochastic Fields

In this section we investigate some properties of the maximum likelihood estimate for drift parameter of a diffusion-type stochastic field. Assume that we observe the generalized diffusion-type stochastic field

$$\xi(z) = \theta\int_{[0,z]} a(u,\xi)du + \int_{[0,z]} b(u,\xi)W(du), \tag{2.36}$$

and we need to estimate the unknown parameter θ, basing on the given observations of $\xi(z)$, $z \in [0,T]^2$. Together with the stochastic field $\xi(z)$ we also observe the random field

$$\eta(z) = \int_{[0,z]} b(u,\xi)W(du). \tag{2.37}$$

Suppose that conditions below are fulfilled:

D1. $P_\theta\left\{\int_{[0,T]^2}\left(\overline{b}^{-1}(z,\xi)a(z,\xi)\right)^2 dz < \infty\right\} = 1,$

D2. $P_\theta\left\{\int_{[0,T]^2}\left(\overline{b}^{-1}(z,\eta)a(z,\eta)\right)^2 dz < \infty\right\} = 1,$

where the index θ means that the distribution of the stochastic field is observed as a function of the parameter θ. We denote by E_θ the expectation in the probability space $(\Omega, \Im, P_\theta)$. Under these conditions we can define the integral $\int_{[0,T]^2}\left(\overline{b}^{-1}(z,\xi)\right)^2 a(z,\xi)\,\xi(dz)$.

Denote by μ_ξ^θ and μ_η the measures in the space $\left(C[0,T]^2,\mathcal{B}\right)$, which correspond to the random fields ξ and η, respectively. According to Theorem 2.12 these measures are equivalent, and

$$\frac{d\mu_\xi^\theta}{d\mu_\eta} = \exp\left\{\theta\int_{[0,T]^2}\left(\overline{b}^{-1}(z,\xi)\right)^2 a(z,\xi)\,\xi(dz) - \frac{\theta^2}{2}\int_{[0,T]^2}\left(\overline{b}^{-1}(z,\xi)a(z,\xi)\right)^2 dz\right\}.$$

If the condition

D3. $P_\theta\left\{\int_{[0,T]^2}\left(\overline{b}^{-1}(z,\xi)\,a(z,\xi)\right)^2 dz > 0\right\} = 1,$

holds true, then for the unknown parameter θ there exists (see [6]) the maximum likelihood estimate of the form

$$\theta_T(\xi) = \frac{\int_{[0,T]^2}\left(\overline{b}^{-1}(z,\xi)\right)^2 a(z,\xi)\,\xi(dz)}{\int_{[0,T]^2}\left(\overline{b}^{-1}(z,\xi)a(z,\xi)\right)^2 dz}.$$

Now we investigate this estimate more in detail.

Lemma 2.11 *Assume that $\delta(x)$ is a measurable continuous functional on $\left(C[0,T]^2,\mathcal{B}\right)$, such that $P_\theta\{\delta(\xi)=0\} = P_\theta\{\delta(W)=0\} = 0$. Put*

$$\varphi_h(\xi) = \left\{\int_{[0,T]^2}\left(\overline{b}^{-1}(z,\xi)\right)^2 a(z,\xi)\xi(dz) - \theta\int_{[0,T]^2}\left(\overline{b}^{-1}(z,\xi)a(z,\xi)\right)^2 dz\right.$$
$$\left. - h\nu\int_{[0,T]^2}\left(\overline{b}^{-1}(z,\xi)a(z,\xi)\right)^2 dz\right\}^2 - \int_{[0,T]^2}\left(\overline{b}^{-1}(z,\xi)a(z,\xi)\right)^2 dz,$$

where $h \in R$, and the random variable ν satisfies $0 < \nu(\xi) < 1$. If $\sup_{-\infty<\theta<\infty} E_\theta\delta(\xi)\varphi_h(\xi) < \infty$, then the following equality holds true:

$$\lim_{h\to 0}\frac{1}{h}\left(E_{\theta+h}\delta(\xi) - E_\theta\delta(\xi)\right) = E_\theta\delta(\xi) - \int_{[0,T]^2}\overline{b}^{-1}(z,\xi)a(z,\xi)W(dz).$$

Proof By the Taylor formula, we obtain for any $h \in R$

$$\lim_{h\to 0}\frac{1}{h}\left(E_{\theta+h}\delta(\xi)-E_\theta\delta(\xi)\right)$$

$$=\frac{1}{h}\delta(\xi)\left[\exp\left\{h\int_{[0,T]^2}\overline{b}^{-1}(z,\xi)a(z,\xi)d\xi(z)-\theta h\int_{[0,T]^2}\left(\overline{b}^{-1}(z,\xi)a(z,\xi)\right)^2dz\right.\right.$$
$$\left.\left.-\frac{h^2}{2}\int_{[0,T]^2}\left(\overline{b}^{-1}(z,\xi)a(z,\xi)\right)^2dz\right\}-1\right]$$
$$=E_\theta\delta(\xi)\left\{\int_{[0,T]^2}\overline{b}^{-1}(z,\xi)a(z,\xi)\xi(dz)-\theta\int_{[0,T]^2}\left(\overline{b}^{-1}(z,\xi)a(z,\xi)\right)^2dz\right\}$$
$$+\frac{h}{2}E_{\theta+h\nu}\delta(\xi)\times\left[\left(\int_{[0,T]^2}\overline{b}^{-1}(z,\xi)a(z,\xi)\xi(dz)-\theta\int_{[0,T]^2}\overline{b}^{-1}(z,\xi)a(z,\xi)dz\right.\right.$$
$$\left.\left.-\nu h\int_{[0,T]^2}\left(\overline{b}^{-1}(z,\xi)a(z,\xi)\right)^2dz\right)^2-\int_{[0,T]^2}\left(\overline{b}^{-1}(z,\xi)a(z,\xi)\right)^2dz\right],$$

where $\nu=\nu(\xi)\in(0,1)$.

By the Scheffe Theorem [4] the sequence of measures $\mu_\xi^{\theta+h\nu}$ on the space $\left(C[0,T]^2,\mathcal{B}\right)$ converges as $h\to 0$ to the probability measure μ_ξ^θ. Therefore, taking into account the last equation and the condition $\sup_{-\infty<\theta<\infty}E_\theta\delta(\xi)\varphi_h(\xi)<\infty$, we obtain the statement of lemma.

Theorem 2.16 *Assume that conditions D1–D3 hold true, and the stochastic field is such that*

(a) $\int_{[0,T]^2}\left(\overline{b}^{-1}(z,\xi)a(z,\xi)\right)^2dz$ is continuous on $\left(C[0,T]^2,\mathcal{B}\right)$.

(b) $\sup_{-\infty<\theta<\infty}E_\theta\int_{[0,T]^2}\left(\overline{b}^{-1}(z,\xi)a(z,\xi)\right)^8dz<\infty$.

(c) $\sup_{-\infty<\theta<\infty}E_\theta\left(\int_{[0,T]^2}\left(\overline{b}^{-1}(z,\xi)a(z,\xi)\right)^2dz\right)^{-8}<\infty$.

Then the shift $b_T(\theta)$ and standard deviation $B_T(\theta)$ are given by

$$b_T(\theta)=\frac{d}{d\theta}E_\theta\left(\int_{[0,T]^2}\left(\overline{b}^{-1}(z,\xi)a(z,\xi)\right)^2dz\right)^{-1},$$

$$B_T(\theta)=E_\theta\left(\int_{[0,T]^2}\left(\overline{b}^{-1}(z,\xi)a(z,\xi)\right)^2dz\right)^{-1}$$
$$+\frac{d^2}{d\theta^2}E_\theta\left(\int_{[0,T]^2}\left(\overline{b}^{-1}(z,\xi)a(z,\xi)\right)^2dz\right)^{-2}.$$

Proof Since

$$\theta_T(\xi) = \theta + \frac{\displaystyle\int_{[0,T]^2} \overline{b}^{-1}(z,\xi)a(z,\xi)\,W(dz)}{\displaystyle\int_{[0,T]^2} \left(\overline{b}^{-1}(z,\xi)a(z,\xi)\right)^2 dz},$$

we have

$$b_T(\theta) = E_\theta(\theta_T - \theta) = E_\theta \frac{\displaystyle\int_{[0,T]^2} \overline{b}^{-1}(z,\xi)a(z,\xi)\,W(dz)}{\displaystyle\int_{[0,T]^2} \left(\overline{b}^{-1}(z,\xi)a(z,\xi)\right)^2 dz}.$$

We put $\delta(\xi) = \left(\int_{[0,T]^2} \left(\overline{b}^{-1}(z,\xi)a(z,\xi)\right)^2 dz\right)^{-1}$ and estimate $E_\theta\delta(\xi)\varphi_h(\xi)$. Taking into account that

$$\begin{aligned}
\phi_h(\xi) = {} & \left(\int_{[0,T]^2} \overline{b}^{-1}(z,\xi)a(z,\xi)\,dz\right)^2 \\
& - 2h\nu\left(\int_{[0,T]^2} \overline{b}^{-1}(z,\xi)a(z,\xi)\,W(dz)\right)\left(\int_{[0,T]^2} \left(\overline{b}^{-1}(z,\xi)a(z,\xi)\right)^2 dz\right) \\
& + h^2\nu^2\left(\int_{[0,T]^2} \left(\overline{b}^{-1}(z,\xi)a(z,\xi)\right)^2 dz\right)^2 - \int_{[0,T]^2} \left(\overline{b}^{-1}(z,\xi)a(z,\xi)\right)^2 dz,
\end{aligned}$$

and applying to all summand the Cauchy–Bunyakovsky–Schwarz inequality, we obtain

$$\begin{aligned}
E_\theta\delta(\xi)\varphi_h(\xi) \le {} & \left[E_\theta\left(\int_{[0,T]^2} \overline{b}^{-1}(z,\xi)a(z,\xi)W(dz)\right)^4\right]^{1/2} \\
& \times \left[E_\theta\left(\int_{[0,T]^2} \left(\overline{b}^{-1}(z,\xi)a(z,\xi)\right)^2 dz\right)^{-2}\right]^{1/2} \\
& + 2hE_\theta\left(\int_{[0,T]^2} \overline{b}^{-1}(z,\xi)a(z,\xi)W(dz)\right)^2 \\
& + 2h^2E_\theta\left(\int_{[0,T]^2} \left(\overline{b}^{-1}(z,\xi)a(z,\xi)\right)^2 dz\right)^{-2} + 1.
\end{aligned}$$

By Theorem 2.3 we have $\sup_{-\infty<\theta<\infty} E_\theta\delta(\xi)\varphi_h(\xi) < \infty$, and thus we can deduce from Lemma 2.11 that

$$b_T(\theta) = E_\theta + \frac{\int_{[0,T]^2} \overline{b}^{-1}(z,\xi)a(z,\xi)\,W(dz)}{\int_{[0,T]^2} \left(\overline{b}^{-1}(z,\xi)a(z,\xi)\right)^2 dz}$$
$$= \frac{d}{d\theta}E_\theta\left(\int_{[0,T]^2} \left(\overline{b}^{-1}(z,\xi)a(z,\xi)\right)^2 dz\right)^{-1}.$$

Now we estimate the standard deviation. It is easy to see that

$$B_T(\theta) = E_\theta(\theta_T - \theta)^2 = E_\theta\left(\int_{[0,T]^2} \overline{b}^{-1}(z,\xi)a(z,\xi)W(dz)\right)^2$$
$$\times\left(\int_{[0,T]^2} \left(\overline{b}^{-1}(z,\xi)a(z,\xi)\right)^2 dz\right)^{-2},$$

and if we take $\delta_0(\xi) = \delta_1(\xi) - \theta\delta(\xi)$, where

$$\delta_1(\xi) = \left(\int_{[0,T]^2} \left(\overline{b}^{-1}(z,\xi)\right)^2 a(z,\xi)\xi(dz)\right)\left(\int_{[0,T]^2} \left(\overline{b}^{-1}(z,\xi)a(z,\xi)\right)^2 dz\right)^{-2},$$

then the standard variation can be written as

$$B_T(\theta) = E_\theta\delta_0(\xi)\int_{[0,T]^2} \overline{b}^{-1}(z,\xi)a(z,\xi)W(dz) = \frac{d}{d\theta}E_\theta\delta_1(\xi) - \theta\frac{d}{d\theta}E_\theta\delta(\xi).$$

Estimation for $\delta_0(\xi)$ can be performed in the same way as that for $\delta(\xi)$. Indeed,

$$
\begin{aligned}
E_\theta\delta_0(\xi)\varphi_h(\xi) \le & \left[E_\theta\left(\int_{[0,T]^2}\overline{b}^{-1}(z,\xi)a(z,\xi)\,W(dz)\right)^6\right]^{1/2} \\
& \times\left[E_\theta\left(\int_{[0,T]^2}\left(\overline{b}^{-1}(z,\xi)a(z,\xi)\right)^2 dz\right)^{-4}\right]^{1/2} \\
& + 2h\left[E_\theta\left(\int_{[0,T]^2}\overline{b}^{-1}(z,\xi)a(z,\xi)W(dz)\right)^4\right]^{1/2} \\
& \times\left[E_\theta\left(\int_{[0,T]^2}\left(\overline{b}^{-1}(z,\xi)a(z,\xi)\right)^2 dz\right)^{-2}\right]^{1/2} \\
& + 2h^2\left[E_\theta\left(\int_{[0,T]^2}\overline{b}^{-1}(z,\xi)a(z,\xi)\,W(dz)\right)^2\right]^{1/2} \\
& + \left[E_\theta\left(\int_{[0,T]^2}\overline{b}^{-1}(z,\xi)a(z,\xi)W(dz)\right)^2\right]^{1/2} \\
& \times\left[E_\theta\left(\int_{[0,T]^2}\left(\overline{b}^{-1}(z,\xi)a(z,\xi)\right)^2 dz\right)^{-2}\right]^{1/2}.
\end{aligned}
$$

Thus, $\sup_{-\infty<\theta<\infty} E_\theta\delta_0(\xi)\varphi_h(\xi) < \infty$, and the proof is finalized by applying Lemma 2.11. □

Chapter 3
Filtration and Prediction Problems for Stochastic Fields

In this chapter we investigate different models of filtration and prediction for stochastic fields generated by some stochastic differential equations. We deduce stochastic integro-differentiation equations for an optimal in the mean square sense filter. We also suggest different approaches for finding the best linear estimate for a stochastic field basing on its observations in certain domain. Besides, we investigate the duality of the filtration problem and a certain optimal control problem. This chapter is based on the results published in [3, 5, 11, 12, 17, 41–44, 46, 69].

3.1 Filtration Problem for Partly Observed Random Fields

We follow the presentation of the results obtained in [11, 15]. For random processes such a problem was investigated in [54].

Let (η,ξ) be a partly observed stochastic field on the plane, i.e., the component $\eta = (\eta(z), \Im_z)$, $z \in [0,1]^2$, is not observed, and $\xi = (\xi(z), \Im_z)$, $z \in [0,1]^2$ is known. To find the optimal filtration for a partly observed random field (η,ξ) we construct for each $z \in [0,1]^2$ an optimal in the mean square sense $\Im_z$-measurable random function $h(z)$, depending on (η,ξ), using the observations of $\xi(z_1)$, $z_1 \leq z$.

Let $\Im_z^\xi = \sigma\left\{\xi(z_1), z_1 \in [0,1]^2, z_1 \leq z = (t,s)\right\}$ and $\Im_z^{\xi *} = \Im_{(1,s)}^\xi \vee \Im_{(t,1)}^\xi$. If $Eh^2(z) < \infty$, the *a posterior* expectation $\pi(z,h) = E\left(h(z)/\Im_z^{\xi *}\right)$ can be taken as required estimation, but the procedure of finding $\pi(z,\zeta)$ without any additional assumptions about the structure of (h,ξ) is an extremely difficult problem. In what follows we consider the model, in which the observed component is an Ito field, and the function $h = (h(z), \Im_z)$, $z \in [0,1]^2$, which is under estimation, can be written in the form

P.S. Knopov and O.N. Deriyeva, *Estimation and Control Problems for Stochastic Partial Differential Equations*, Springer Optimization and Its Applications 83,
DOI 10.1007/978-1-4614-8286-4_3, © Springer Science+Business Media New York 2013

$$h(z) = h(0) + \int_{[0,z]} H(u)du + X(z). \tag{3.1}$$

Here $X = (X(z), \Im_z)$, $z \in [0,1]^2$ is a strong martingale. Further we assume that the conditions below hold true:

G1. $P\left\{\int_{[0,1]^2} |H(u)|du < \infty\right\} = 1$

G2. $\sup_{[0,1]^2} Eh^2(z) < \infty$.

For any measurable stochastic field $h(z) = h(z,\omega)$ with $E|h(z)| < \infty$ we denote

$$\pi(z,h) = E\left(h(z)/\Im_z^{\xi*}\right)$$

and

$$\varphi(z,h) = E\left(\pi(z,h)/\Im_z^{\xi}\right).$$

We assume that the observed field $\xi = (\xi(z), \Im_z)$ is an Ito field with respect to the Wiener field $W = (W(z), \Im_z)$, i.e.,

$$\xi(z) = \xi(0) + \int_{[0,z]} A(u)du + \int_{[0,z]} B(u,\xi)W(du).$$

In addition, we assume that the random fields A and B satisfy the conditions

G3. $P\left\{\int_{[0,1]^2} |A(u)|du < \infty\right\} = 1, P\left\{\int_{[0,1]^2} B^2(u,\xi)du < \infty\right\} = 1;$

G4. the functional $B(z,x)$ is $\mathcal{B}_z$-measurable for any $x \in C[0,1]^2$ (where $\mathcal{B}_z$ is defined in Sect. 2.1);

G5. $|B(z,x)| \leq k(1 + \|x\|_z)$ for some constant value k;

G6. $|B(z,x) - B(z,x_1)| \leq c\|x - x_1\|_z$ for some constant value c;

G7. $P\left\{\int_{[0,1]^2} \left(\overline{B}^{-1}(u,\xi)A(u)\right)^2 du < \infty\right\} = 1$ (where $\overline{f}^{-1}(z)$ is defined in Sect. 1.1)

Let the field $\hat{W} = \left(\hat{W}(z), \Im_z^{\xi}\right)$ be given by

$$\hat{W}(z) = \int_{[0,z]} \overline{B}^{-1}(u,\xi)\xi(du) - \int_{[0,z]} \overline{B}^{-1}(u,\xi)\varphi(u,A)du.$$

It follows from Theorem 2.1 that it is the Wiener field on $[0,1]^2$, and ξ can be represented in the form

$$\xi(z) = \xi(0) + \int_{[0,z]} \phi(u, A)du + \int_{[0,z]} B(u, \xi)\hat{W}(du). \tag{3.2}$$

Theorem 3.1 *Suppose that the partly observed stochastic field (h,ξ) can be represented as in (3.1) and (3.2), and the random fields H, X, A, B satisfy conditions G1, G3–G7. Then for any* $z \in [0,1]^2$

$$\pi(z, h) = \pi(0, h) + \int_{[0,z]} \pi(u, H)du + \int_{[0,z]} \pi(u, G)\hat{W}(du)$$

$$+ \int_{[0,z]} \overline{B}^{-1}(u, \xi)(\pi(u, hA) - \pi(u, h)\phi(u, A))\hat{W}(du),$$

with probability 1, and the stochastic field $G = (G(z), \Im_z)$ *is such that* $G(z)dz = d<X, W>_z$.

Now we will prove some auxiliary propositions needed to check the proposition of Theorem 3.1.

Lemma 3.1 *Under the condition G2 stochastic field* $\left(E\left[h(0)/\Im_z^{\xi*}\right], \Im_z\right)$ *is square integrable strong martingale admitting the representation*

$$E\left[h(0)/\Im_z^{\xi*}\right] = \pi(0, h) + \int_{[0,z]} g^h(u, \xi)\hat{W}(du),$$

where $g^h(z,\xi)$, $z \in [0,1]^2$, *is some stochastic field, satisfying*

$$E \int_{[0,1]^2} \left(g^h(u, \xi)\right)^2 du < \infty.$$

Proof For any $z_1 < z$ we have

$$E\left(E\left[h(0)/\Im_z^*\right]/\Im_{z_1}^*\right) = E\left[h(0)/\Im_{z_1}^*\right],$$

Therefore $E\left[h(0)/\Im_z^{\xi*}\right]$ is a strong martingale, which has right limits with probability 1 for any $z \in [0,1]^2$. Moreover, applying Theorem 1.12 we obtain

$$E\left[h(0)/\Im_z^{\xi*}\right] = m_0 + \int_{[0,z]} g^h(u, \xi)\hat{W}(du),$$

with

$$m_0 = E\left[h(0)/\Im_0^{\xi*}\right] = \pi(0, h),$$

$$E\int\limits_{[0,1]^2} \left(g^h(u,\xi)\right)^2 du < \infty. \qquad \square$$

Lemma 3.2 *Under condition G2, the random field* $\left(E\left[X(z)/\Im_z^{\xi*}\right], \Im_z\right)$ *is the square integrable martingale, possessing the representation*

$$E\left[X(z)/\Im_z^{\xi*}\right] = \int\limits_{[0,z]} g^X(u,\xi)\hat{W}(du),$$

for some $g^X(z,\xi)$ *such that*

$$E\int\limits_{[0,1]^2} \left(g^X(u,\xi)\right)^2 du < \infty.$$

Proof Put $y(z) = E\left[X(z)/\Im_z^{\xi*}\right]$. By the Jensen inequality we derive

$$E|y(z)| \le E\left|E\left(X(z)/\Im_z^{\xi*}\right)\right| \le E|X(z)|, z\in[0,1]^2.$$

For any $z_1 < z$ we obtain with probability 1

$$E\left(y(z_1,z]/\Im_z^*\right) = E\left(E\left[X(z_1,z]/\Im_{z_1}^{\xi*}\right]/\Im_{z_1}^*\right) = 0.$$

Moreover, since $X(z)$ equals to zero on both axes, we have $y(0,s) = y(t,0) = 0$. Therefore, $y = (y(z), \Im_z)$ is the strong martingale, implying that

$$E\left[X(z)/\Im_z^{\xi*}\right] = \int\limits_{[0,z]} g^X(u,\xi)\hat{W}(du),\ E\int\limits_{[0,1]^2} \left(g^X(u,\xi)\right)^2 du < \infty. \qquad \square$$

Lemma 3.3 *Assume that condition G2 is fulfilled. Then the stochastic field* $\hat{y} = (\hat{y}(z), \Im_z)$ *given by the equation*

$$\hat{y}(z) = E\left[\int\limits_{[0,z]} H(u)du/\Im_z^{\xi*}\right] - \int\limits_{[0,z]} \pi(u,H)du,$$

is a square integrable strong martingale, admitting the representation

$$\hat{y}(z) = \int\limits_{[0,z]} g^H(u,\xi)\hat{W}(du), E\int\limits_{[0,1]^2} \left(g^H(u,\xi)\right)^2 du < \infty.$$

Proof Since $\pi(z,H) = E\left[H(z)/\Im_z^{\xi*}\right]$, the equality

$$E\left[\int\limits_{[0,z]} H(u)du/\wp\right] = \int\limits_{[0,z]} E\left[H(u)/\wp\right]du \tag{3.3}$$

holds true for any σ-algebra $\wp \subset \Im$, which in turn gives

$$\hat{y}(z) = \int\limits_{[0,z]} E\left[H(u)/\Im_z^{\xi*}\right]du - \int\limits_{[0,z]} E\left[H(u)/\Im_u^{\xi*}\right]du.$$

Thus, for any $z_1 < z$ we obtain

$$\hat{y}(z_1,z] = \int\limits_{[z_1,z]} E\left[H(u)/\Im_z^{\xi*}\right]du - \int\limits_{[z_1,z]} E\left[H(u)/\Im_u^{\xi*}\right]du.$$

Applying (3.3) to $E\left[y(z_1,z]/\Im_{z_1}^{\xi*}\right]$ we derive

$$\begin{aligned} E\left[\hat{y}(z_1,z]/\Im_{z_1}^*\right] &= \int\limits_{[z_1,z]} E\left(E\left[H(u)/\Im_z^{\xi*}\right]/\Im_{z_1}^*\right)du - \int\limits_{[z_1,z]} E\left(E\left[H(u)/\Im_u^{\xi*}\right]/\Im_{z_1}^*\right)du \\ &= 0. \end{aligned}$$

Moreover, since $\hat{y}(0,s) = \hat{y}(t,0) = 0$, the field $\hat{y} = (\hat{y}(z), \Im_z)$ is the strong martingale and, consequently, can be written as

$$\hat{y}(z) = \int\limits_{[0,z]} g^H(u,\xi)\hat{W}(du), E\int\limits_{[0,1]^2} \left(g^H(u,\xi)\right)^2 du < \infty. \qquad \square$$

Proof of the Theorem 3.1 From Lemmas 3.1–3.3 we have

$$\pi(z,h) = \pi(0,h) + \int\limits_{[0,z]} \pi(u\hat{W},H)du + \int\limits_{[0,z]} g(u,\xi)\hat{W}(du),$$

$$g(z,\xi) = g^h(z,\xi) + g^X(z,\xi) + g^H(z,\xi),$$

$$E\int\limits_{[0,1]^2} g(u,\xi)du < \infty.$$

Let $\lambda = \left(\lambda(z,\xi), \Im_z^\xi\right)$ be a bounded random field. Denote

$$y(z) = \int_{[0,z]} g(u,\xi)\hat{W}(du), \zeta(z) = \int_{[0,z]} \lambda(u)\hat{W}(du).$$

Then obviously

$$Ey(z)\lambda(z) = E\int_{[0,z]} \lambda(u,\xi)g(u,\xi)\hat{W}(du). \tag{3.4}$$

On the other hand, we can write $y(z)$ in the form

$$y(z) = \pi(z,h) - \pi(0,h) - \int_{[0,z]} \pi(u,H)du,$$

$$E\zeta(z)\pi(0,H) = E\left[\pi(0,H)E\left(\zeta(z)/\Im_0^\xi\right)\right] = 0.$$

Therefore $Ey(z)\lambda(z)$ can be calculated as follows:

$$E\left(\zeta(z)\int_{[0,z]} \pi(u,H)du\right) = \int_{[0,z]} E(\zeta(z)\pi(u,H))du = \int_{[0,z]} E\left(E\left[\zeta(z)/\Im_u^{\xi*}\right]\pi(u,H)\right)du$$
$$= \int_{[0,z]} E(\zeta(u)\pi(u,H))du.$$

Since the field $\zeta(z)$ is $\Im_z^{\xi*}$-measurable, we get

$$Ey(z)\zeta(z) = E\zeta(z)\pi(z,h) - \int_{[0,z]} \zeta(u)\pi(u,H)du$$
$$= E\left(\zeta(z)E\left[\pi(z,h)/\Im_z^{\xi*}\right]\right) - \int_{[0,z]} E\left(\zeta(u)E\left[H(u)/\Im_z^{\xi*}\right]\right)du$$
$$= E\left(\zeta(z)h(z) - \int_{[0,z]} \zeta(u)H(u)du\right).$$

Further,

$$\hat{W}(z) = W(z) + \int\limits_{[0,z]} \overline{B}^{-1}(u,\xi)(A(u) - \varphi(u,A))du,$$

which implies

$$\zeta(z) = \int\limits_{[0,z]} \lambda(u,\xi)W(du) + \int\limits_{[0,z]} \lambda(u,\xi)\overline{B}^{-1}(u,\xi)(A(u) - \varphi(u,A))du.$$

Put $\zeta^*(z) = \int_{[0,z]}\lambda(u,\xi)W(du)$. Then

$$\begin{aligned} Ey(z)\zeta(z) =& E\left(\zeta^*(z)h(z) - \int\limits_{[0,z]} \zeta^*(u)H(u)du\right) \\ &+ E\left(h(z)\int\limits_{[0,z]} \lambda(u,\xi)\overline{B}^{-1}(u,\xi)(A(u) - \varphi(u,A))du\right. \\ &\left.- \int\limits_{[0,z]}\int\limits_{[0,u]} \left(\lambda(v,\xi)\overline{B}^{-1}(v,\xi)(A(v) - \varphi(v,A))dv\right)H(u)du\right). \end{aligned} \quad (3.5)$$

Since $\zeta^* = (\zeta^*(z), \Im_z)$ is a square integrable strong martingale, we get

$$E\zeta^*(z)h(0) = E(h(0)E[\zeta^*(z)/\Im_0]) = Eh(0)\zeta^*(0) = 0.$$

Moreover,

$$E\int\limits_{[0,z]} \zeta^*(u)H(u)du = E\int\limits_{[0,z]} E[\zeta^*(u)/\Im_u^*]H(u)du,$$

implying

$$\begin{aligned} E\left(\zeta^*(z)h(z) - \int\limits_{[0,z]} \zeta^*(u)H(u)du\right) &= E\left(\zeta^*(z)\left[h(z) - h(0) - \int\limits_{[0,z]} H(u)du\right]\right) \\ &= E\zeta^*(z)X(z) = E < \zeta^*, X>_z, \end{aligned}$$

and therefore $< \zeta^*, X >_z = \int_{[0,z]}\lambda(u,\xi)d < X, W >_u$. Hence,

$$E\left(\zeta^*(z)h(z) - \int\limits_{[0,z]} \zeta^*(u)H(u)du\right) = E\left(\int\limits_{[0,z]} \lambda(u,\xi)d <X,W>_u\right)$$
$$= E\left(\int\limits_{[0,z]} \lambda(u,\xi)\pi(u,G)du\right).$$

Let us calculate now the second summand in (3.5). We have

$$E\left(h(z)\int\limits_{[0,z]} \lambda(u,\xi)\overline{B}^{-1}(u,\xi)[A(u)-\varphi(u,A)]du\right)$$
$$= E\left(\int\limits_{[0,z]} \lambda(u,\xi)h(u)\overline{B}^{-1}(u,\xi)[A(u)-\varphi(u,A)]du\right)$$
$$+ E\left(\int\limits_{[0,z]} \lambda(u,\xi)[h(z)-h(u)]\overline{B}^{-1}(u,\xi)[A(u)-\varphi(u,A)]du\right)$$
$$= E\left(\int\limits_{[0,z]} \lambda(u,\xi)\overline{B}^{-1}(u,\xi)[\pi(u,hA)-\pi(u,h)\varphi(u,A)]du\right)$$
$$+ E\left(\int\limits_{[0,z]} \lambda(u,\xi)[h(z)-h(u)]\overline{B}^{-1}(u,\xi)[A(u)-\varphi(u,A)]du\right).$$

Since

$$h(z)-h(u) = X(z)-X(u) + \int\limits_{[u,z]} H(v)dv,$$

and X is a strong martingale, we obtain

$$E\left(\int\limits_{[0,z]} \lambda(u,\xi)[h(z)-h(u)]\overline{B}^{-1}(u,\xi)[A(u)-\varphi(u,A)]du\right)$$
$$= E\left(\int\limits_{[0,z]} H(u)\int\limits_{[0,u]} \lambda(v,\xi)\overline{B}^{-1}(u,\xi)[A(u)-\varphi(u,A)]dvdu\right).$$

From the last formula we derive

$$E\left(h(z)\int_{[0,z]}\lambda(u,\xi)\overline{B}^{-1}(u,\xi)[A(u)-\varphi(u,A)]du\right)$$

$$=E\left(\int_{[0,z]}\lambda(u,\xi)\overline{B}^{-1}(u,\xi)[\pi(u,hA)-\pi(u,h)\varphi(u,A)]du\right)$$

$$+E\left(\int_{[0,z]}H(u)\int_{[0,u]}\lambda(v,\xi)\overline{B}^{-1}(u,\xi)[A(u)-\varphi(u,A)]dvdu\right)$$

Therefore, we obtained the representation

$$Ey(z)\zeta(z)=E\left(\int_{[0,z]}\lambda(u,\xi)\pi(u,G)du\right)$$

$$+E\left(\int_{[0,z]}\lambda(u,\xi)\overline{B}^{-1}(u,\xi)[\pi(u,hA)-\pi(u,h)\varphi(u,A)]du\right).$$

Recall that $\int_{[0,z]} h(z,\xi)\hat{W}(du)$ does not change if we modify $h(z,\xi)$ on the set of Lebesgue measure 0. Comparing the last equation with (3.4) we obtain

$$\pi(z,h)=\pi(0,h)+\int_{[0,z]}\pi(u,H)du+\int_{[0,z]}\pi(u,G)\hat{W}(du)$$

$$+\int_{[0,z]}\overline{B}^{-1}(u,\xi)(\pi(u,hA)-\pi(u,h)\varphi(u,A))\hat{W}(du).\qquad\square$$

Consider now an important particular case of the theorem proved above, namely, when $A(z)\equiv 0$, $B(z)\equiv 1$, $\xi(0)=0$.

Corollary 3.1 Let $W=(W(z),\Im_z)$, $z\in[0,1]^2$ be a Wiener field, and let the stochastic field h be given by (3.1). If conditions D1–D2 are satisfied, then

$$\pi^W(z,h)=\pi^W(0,h)+\int_{[0,z]}\pi^W(u,H)du+\int_{[0,z]}\pi^W(u,G)W(du),$$

where $\pi^W(z,h)=E\left[h(z)/\Im_z^{W*}\right]$ and random field $G=(G(z),\Im_z)$ is such that $G(z)\,dz=d\,\langle X,W\rangle_z$.

The proof follows directly from the fact that $\xi(z)=W(z)=\hat{W}(z)$.

3.2 Filtration Problem for Stochastic Fields Observed in the Half-Space

In this section we follow the results presented in [41]. Consider on the probability space $(\Omega, \Im, P)$ a stochastic field $X(z,x)$ generated by the stochastic differential equation

$$\frac{\partial X(z,x)}{\partial x} = L_z X(z,x) + K_z \dot{W}(z,x), z \in R^2, x \in R, \tag{3.6}$$

where $W(\Delta)$ is the Gaussian orthogonal stochastic measure, defined R^3 with $EW(\Delta) = 0$ and $E|d, W(z,x)|^2 = dzdx$. In what follows we denote by $\dot{W}(z,x)$ the derivative of the orthogonal measure $W(\Delta)$. Clearly, $\dot{W}(z,x)$ does not exist in the common sense and is to be understood as a generalized random field, often called the *white noise*. Here L_z and K_z are the partial differential operators acting on z, with coefficients depending on (z, x). We assume that the coefficients of operators L_z and K_z are smooth enough in z and continuous with respect to x. We also assume that the orders of differentiation in the operator K_z with respect to each variable of the operator is lower than those in L_z. Let l be the highest order of the derivative in L_z.

We denote by $S(D)$, $D \subset E_{n+1}$, the class of infinitely many times differentiable with respect to (z, x) functions $f(z,x)$, which are equal to zero on the boundary of D and outside D. Finally, we denote by L_z^* and K_z^* the adjoint operators to L_z and K_z, respectively. Finally, denote by L_z^* and K_z^* the adjoint operators to L_z and K_z, respectively.

Definition 3.1 We say that $X(z,x)$ is the solution to (3.6), if for any $A(z,x) \in S_l(D)$ and any domain D the equality

$$\int_D L_z^*(A(z,x))X(z,x)dzdx = \int_D K_z^*(A(z,x))dW(z,x)$$

holds true with probability 1. The integral in the left-hand side of the formula is understood as the ordinary stochastic integral with respect to the orthogonal measure $W(\Delta)$.

Remark 3.1 The class of functions $A(z,x)$ considered here is more narrower than one can demand. It is enough to demand that $A(z,x)$ belongs to the domain of the operator $\frac{\partial}{\partial x} - L_z$, but for simplicity we restrict ourselves to the class $S_l(D)$.

Further, we assume the sufficient conditions for existence of the fundamental solution to the equation $\frac{\partial X(z)}{\partial x} = L_z X(z,x)$ to hold true.

Theorem 3.2 [41] *Consider the random field*

$$X(z,x) = \int_{-\infty}^{x} \int_{R^2} K_{z'}^{*} G(z,x,z',\tau) dW(z',\tau)$$

where $G(z,x,z',y)$ is the fundamental solution to the equation

$$\frac{\partial X(z)}{\partial x} = L_z X(z,x).$$

Then the stochastic field $X(z,x)$ is (with probability 1) the solution to (3.6) in the domain $R^{n+1} = R^n \times (-\infty, +\infty)$, and the correlation function of $X(z,x)$ is

$$R(z_1,z_2,x_1,x_2) = \int_{-\infty}^{\min(x_1,x_2)} \int_{R^2} K_{z'}^{*} G(z_1,x_1,z',\tau) K_{z'}^{*} G(z_2,x_2,z',\tau) dz' d\tau.$$

Note that the equations $L_t u(t) = K_t \dot{W}(t)$, where L_t, K_t are ordinary differential operators, are comprehensively studied in [16].

Now we turn to problem of estimation of the random field $X(z,x)$.

Suppose that we observe in the domain $\Omega \times [0, x)$ a random function $Y(z,y)$, related to $X(z,x)$ by

$$dY(z,y) = H(z,y)X(z,y)dzdy + \sqrt{r(z,y)}dV(z,y), \tag{3.7}$$

where r(z,y) > 0, and $V(\Delta)$ is an orthogonal stochastic measure defined on R^3, such that $EV(\Delta) = 0$ *and* $E|dV(z,x)|^2 = dzdx$.

Here the above equation is understood in the sense that

$$\int_0^{z_1} \int_{\Omega_1} b(z,y) dY(z,y) = \int_0^{z_1} \int_{\Omega_1} b(z,y) H(z,y) X(z,y) dzdy + \int_0^{z_1} \int_{\Omega_1} b(z,y) \sqrt{r(z,y)} dV(z,y) \tag{3.8}$$

holds for any $b(z,y)$, which is square integrable on $[0,z_1] \times \Omega_1$, and $\Omega_1 \subseteq \Omega$. The second term in the right-hand side of (3.8) is the stochastic integral with respect to the orthogonal measure. In what follows we use the conventional representation of (3.7) as $\dot{Y}(z,y) = H(z,y)X(z,y) + \sqrt{r(z,y)}\ \dot{V}(z,y)$, by which we mean that $Y(z,y)$ satisfies (3.8), where $\dot{V}(z,x)$ is the white noise.

We assume that the measures $W(\Delta_1)$ and $V(\Delta_2)$ are mutually uncorrelated, i.e., $EW(\Delta_1)V(\Delta_2) = 0$ for any Δ_1 and Δ_2. The problem we will deal with is the following: Find an estimate $\breve{X}(z,x)$ of the random field $X(z,x)$ by the given observations of the field $Y(z,y)$ in the domain $\Omega \times [0,x]$, such that the minimal possible value of the least square error is attained on $\breve{X}(z,x)$, i.e.,

$$E\left|\breve{X}(z,x) - X(z,x)\right|^2 \to \min.$$

We are looking for the estimate in the form

$$\breve{X}(z,x) = \int_0^x \int_\Omega A(z,x,z',x')dY(z',x').$$

The function $A(z,x,z',x')$ is assumed to be bounded, continuously differentiable in x and x' if $x' \leq x$, and l times differentiable with respect to z and z', $z \neq z'$. It is not hard to deduce the integral equation which the weight function $A(z,x,z',x')$ giving the minimum of the mean squares estimate is due to satisfy. Using our conventional representation, this equation can be written as

$$EX(z,x)\ \dot{Y}(z',x') = \int_0^x \int_\Omega A(z,x,\sigma,\upsilon)E\dot{Y}(\sigma,\upsilon)\dot{Y}(z',x')d\sigma d\upsilon, \tag{3.9}$$

which in turn implies that the equality below

$$\int_0^{x_1} \int_\Omega b(z',x')EX(z,x)\ \dot{Y}(z',x')$$

$$= \int_0^{x_1} \int_\Omega b(z',x')E\int_0^{x_1} \int_\Omega A(z,x,\sigma,\upsilon)dY(\sigma,\upsilon)dY(z',x') \tag{3.10}$$

holds true for any function $b(z,x)$, square integrable on $[0,x_1] \times \Omega_1$, where $x_1 \leq x$, and $\Omega_1 \subseteq \Omega$. This equation can be easily derived by using variation methods.

From (3.9) and (3.10) we obtain the equation for the optimal filter, which in analogy with the one-dimensional case we call the Kalman equation. The filter itself, obtained as the solution to this equation, we call the optimal Kalman filter. In what follows we use the conventional notation (3.9), always having in mind that it is understood in the sense of (3.10). From (3.9) we obtain

$$\frac{\partial}{\partial x}EX(z,x)\ \dot{Y}(z',x') = E\left[L_zX(z,x)\ \dot{Y}(z',x') + K_z\dot{W}(z,x)\dot{Y}(z',x')\right],$$

and differentiating the right-hand side of (3.9) with respect to x we get

$$\frac{\partial}{\partial x}\int_0^x\int_\Omega A(z,x,\sigma,\upsilon)E\dot{Y}(\sigma,\upsilon)\ \dot{Y}(z',x')d\sigma d\upsilon$$
$$=\int_0^x\int_\Omega \frac{\partial A(z,x,\sigma,\upsilon)}{\partial x}E\dot{Y}(\sigma,\upsilon)\ \dot{Y}(z',x')d\sigma d\upsilon$$
$$+\int_\Omega A(z,x,\sigma,\upsilon)E\dot{Y}(\sigma,\upsilon)\ \dot{Y}(z',x')d\sigma d\upsilon.$$

It is easy to see that for $x' < x$ one has

$$E\dot{Y}(\sigma,x)\ \dot{Y}(z',x') = H(\sigma,x)EX(\sigma,x)\ \dot{Y}(z',x').$$

Combining these equations we obtain

$$L_z A(z,x,\sigma,\upsilon) - \frac{\partial A(z,x,\sigma,\upsilon)}{\partial x} - \int_\Omega A(z,x,\alpha,\upsilon)H(\alpha,x)A(\alpha,x,\sigma,\upsilon)d\alpha = 0, \quad (3.11)$$

and if the function $A(z,x,z',x')$ satisfies (3.11), it also satisfies (3.9). Using the last equation, let us derive the equation for the optimal filter. We have

$$\frac{\partial \breve{X}(z,x)}{\partial x} = \int_0^x\int_\Omega \frac{\partial A(z,x,\sigma,\upsilon)}{\partial x}\ \dot{Y}(\sigma,\upsilon)d\sigma d\upsilon + \int_\Omega A(z,x,\sigma,x)\ \dot{Y}(\sigma,x)d\sigma$$
$$=\int_0^x\int_\Omega \left[L_x A(z,x,\sigma,\upsilon) - \int_\Omega A(z,x,\alpha,\upsilon)H(\alpha,x)A(\alpha,x,\sigma,\upsilon)d\alpha\right]\ \dot{Y}(\sigma,\upsilon)d\sigma d\upsilon$$
$$+\int_\Omega A(z,x,\sigma,x)\ \dot{Y}(\sigma,x)d\sigma$$
$$= L_z\,\breve{X}(z,x) - \int_\Omega A(z,x,\sigma,x)H(\sigma,x)\ \breve{X}(\sigma,x)d\sigma + \int_\Omega A(z,x,\sigma,x)\ \dot{Y}(\sigma,x)d\sigma$$
$$= L_z\,\breve{X}(z,x) + \int_\Omega A(z,x,\sigma,x)\left[\dot{Y}(\sigma,x) - H(\sigma,x)\ \breve{X}(\sigma,x)\right]d\sigma.$$

In such a way,

$$\frac{\partial \breve{X}(z,x)}{\partial x} = L_z\,\breve{X}(z,x) + \int_\Omega A(z,x,\sigma,x)\left[\dot{Y}(\sigma,x) - H(\sigma,x)\ \breve{X}(\sigma,x)\right]d\sigma,$$

which, taking into account (3.10), is equivalent to

$$\breve{X}(z,x) = X(z,0) + \int_0^x L_z \breve{X}(z,y)dy$$

$$+ \int_0^x \int_\Omega A(z,y,\sigma,y)dY(\sigma,y) - \int_0^x \int_\Omega H(\sigma,y)X(\sigma,y)d\sigma dy. \quad (3.12)$$

In this equation for the optimal filter we still have one unknown function $A(z,x,\sigma,x) = k(z,\sigma,x)$. To find it, put

$$\widetilde{X}(z,x) = X(z,x) - \breve{X}(z,x),$$

which satisfies the following integro-differential equation

$$\frac{\partial \widetilde{X}(z,x)}{\partial x} = L_z \widetilde{X}(z,x) - \int_\Omega k(z,\sigma,x)H(\sigma,x)\widetilde{X}(\sigma,x)d\sigma + K_z\ \dot{W}(z,x)$$

$$- \int_\Omega k(z,\sigma,x)\ \dot{V}(\sigma,x)d\sigma. \quad (3.13)$$

Although (3.11) understood in the generalized sense, its solution is a stochastic field, because $\widetilde{X}(z,x)$ and $\breve{X}(z,x)$ are stochastic fields. Note that (3.13) contains the function $k(z,\sigma,x)$ we are looking for. Using this function, it is easy to get

$$E\widetilde{X}(z,x)\,\breve{X}(z',x) = 0 \ \text{ if } \ z \neq z',$$

and thus

$$k(z,z',x) = \frac{p(z,z',x)H(z',x)}{r(z',x)},$$

where $p(z,z',x) = E\widetilde{X}(z,x)\widetilde{X}(z',x)$. In such a way, the problem is reduced to deriving $p(z,z',x)$ from the equation

$$\frac{\partial p(z,z',x)}{\partial x} = E\left(\frac{\partial \widetilde{X}(z,x)}{\partial x}\widetilde{X}(z',x)\right) + E\left(\widetilde{X}(z,x)\frac{\partial \widetilde{X}(z',x)}{\partial x}\right).$$

For the first term we have

$$E\left(\frac{\partial \widetilde{X}(z,x)}{\partial x}\widetilde{X}(z',x)\right) = E\left\{L_z\widetilde{X}(z,x) - \int_\Omega k(z,\sigma,x)H(\sigma,x)\widetilde{X}(\sigma,x)d\sigma\right\}\widetilde{X}(\sigma,x)$$

$$+ E\left\{K_z\ \dot{W}(z,x) - \int_\Omega k(z,\sigma,x)\sqrt{r(\sigma,x)}\ \dot{V}(\sigma,x)d\sigma\right\}\widetilde{X}(\sigma,x).$$

After easy transformation one can get

$$E\left(\frac{\partial \widetilde{X}(z,x)}{\partial x}\widetilde{X}(z',x)\right) = L_z p(z,z',x) - \int_{\Omega} k(z,\sigma,x)H(\sigma,x)p(\sigma,z',x)d\sigma$$
$$+\frac{1}{2}\int_{\Omega} k(z,\sigma,x)r(\sigma,x)k(z',\sigma,x)d\sigma$$
$$= L_z p(z,z',x) - \frac{1}{2}\int_{\Omega} \frac{p(z,\sigma,x)p(z',\sigma,x)h^2(\sigma,x)}{r(\sigma,x)}d\sigma.$$

Similarly,

$$E\left(\widetilde{X}(z,x)\frac{\partial \widetilde{X}(z',x)}{\partial x}\right) = L_{z'} p(z,z',x) - \frac{1}{2}\int_{\Omega} \frac{p(z,\sigma,x)p(z',\sigma,x)h^2(\sigma,x)}{r(\sigma,x)}d\sigma.$$

Combining the above calculations, we derive for $p(z,z',x)$ the following equation:

$$\frac{\partial p(z,z',x)}{\partial x} = L_z p(z,z',x) + L_{z'} p(z,z',x) - \int_{\Omega} \frac{p(z,\sigma,x)p(z',\sigma,x)h^2(\sigma,x)}{r(\sigma,x)}d\sigma.$$

To solve it we need to pose the initial conditions on $p(z,z',0)$. Since $\widetilde{X}(z,0) = X(z,0)$, we have

$$p(z,z',0) = EX(z,0)X(z',0) = \int_{-\infty}^{0} K_\sigma^* G(z,0,\sigma,\tau)K_\sigma^* G(z',0,\sigma,\tau)d\sigma d\tau \quad \square$$

Let us briefly dwell on the prediction problem, related to the filtration problem considered above. Suppose that we need to estimate the stochastic field $u(z,x)$ in the point z, basing on the observations of the field $\varsigma(z,x)$ in the domain $[0,z_1] \times \Omega$, where $z_1 < z$. Then the integral equation for the optimal prediction takes the form

$$Eu(z,x)\ \dot{\varsigma}(z',y) = \int_{[0,z_1]}\int_{\Omega} a(z,x,\sigma,\nu)E\ \dot{\varsigma}(\sigma,\nu)\ \dot{\varsigma}(z',y)d\sigma d\nu.$$

Differentiating both sides with respect to x, we get

$$\frac{\partial}{\partial x}Eu(z,x)\ \dot{\varsigma}(z',y) = E\Big[L_z u(z,x)\ \dot{\varsigma}(z',y) + K_z\ \dot{W}(z,x)\ \dot{\varsigma}(z',y)\Big],$$

$$\frac{\partial}{\partial x}\int\limits_{[0,z_1]}\int\limits_{\Omega} a(z,x,\sigma,\nu)E\ \dot{\varsigma}(\sigma,\nu)\ \dot{\varsigma}(z',y)d\sigma d\nu$$
$$=\int\limits_{[0,z_1]}\int\limits_{\Omega}\frac{\partial a(z,x,\sigma,\nu)}{\partial x}E\ \dot{\varsigma}(\sigma,\nu)\ \dot{\varsigma}(z',y)d\sigma d\nu.$$

Consequently,

$$L_z a(z,x,\sigma,\nu)-\frac{\partial a(z,x,\sigma,\nu)}{\partial x}=0, \sigma\le z_1, \nu\in\Omega.$$

It is also possible to show the converse. We can see from the last equation that for the optimal estimate $\breve{u}(z,x)$ of the field $u(z,x)$ one has

$$\frac{\partial\,\breve{u}(z,x)}{\partial x}=\int\limits_{[0,z_1]}\int\limits_{\Omega}\frac{\partial a(z,x,\sigma,\nu)}{\partial x}\ \dot{\varsigma}(\sigma,\nu)d\sigma d\nu=\int\limits_{[0,z_1]}\int\limits_{\Omega}L_z a(z,x,\sigma,\nu)\ \dot{\varsigma}(\sigma,\nu)d\sigma d\nu$$
$$=L_z\ \breve{u}(z,x).$$

Therefore, we derive the equation

$$\frac{\partial\,\breve{u}(z,x)}{\partial x}=\int\limits_{[0,z_1]}\int\limits_{\Omega}\frac{\partial a(z,x,\sigma,\nu)}{\partial x}\ \dot{\varsigma}(\sigma,\nu)d\sigma d\nu=L_z\ \breve{u}(z,x).$$

The initial condition is defined by the estimate of the field $\breve{u}(z,x)$ in the point $z=z_1$. Namely, to determine the initial condition we need to solve the optimal filter equation.

3.3 Innovation Method for Filtration Problems

In this section we follow the results presented in [3]. We have used the idea of the "innovation" method for filtration problem of the random processes [37]. Assume that we have on the probability space $(\Omega,\Im,P)$ a stochastic field generated by the stochastic differential equation

$$\frac{\partial X(z,x)}{\partial x}=L_z X(z,x)+K_z\ \dot{W}(z,x), z\in R^2, x\in[0,+\infty), \tag{3.14}$$

with initial condition $X(z,0)=\Phi(z)$. Here $W(\Delta)$ is the Gaussian orthogonal stochastic measure defined on $R^2\times[0,+\infty)$ with $EW(\Delta)=0$, with structure function $\mu(\Delta)$, which is the a ordinary Lebesgue measure on $R\times[0,+\infty)$ (see, for instance, [28]). As before, we denote by $\dot{W}$ the derivative of the orthogonal measure W. Here L_z and K_z are the partial differential operators acting on z, with coefficients depending on (z, x).

We understand the solution to (3.14) in the generalized sense. As before, denote by $G(z,x,z',y)$ the fundamental solution to the equation $\frac{\partial X(z)}{\partial x} = L_z X(z,x)$. Then (see [41]) the stochastic field $X(z,x)$ defined by

$$X(z,x) = \int_{R^2} G(z,x,z',0)\Phi(z')dz' + \int_0^x \int_{R^2} K^*_{z'} G(z,x,z',\tau)dW(z',\tau), \tag{3.15}$$

solves with probability 1 equation (3.14).

Suppose that we observe in the region $[0,T]^2 \times [0,x]$ a random function $Y(z,y)$, which is related to the random field $X(z,x)$ as follows:

$$dY(z,y) = H(z,y)X(z,y)dzdy + dV(z,y),$$

where $V(\Delta)$ is the Gaussian orthogonal stochastic measure defined on $R^2 \times [0, +\infty)$, with zero expectation and the structure function $\mu(\Delta)$. Measures $W(\Delta_1)$ and $V(\Delta_2)$ are orthogonal for any $\Delta_1, \Delta_2 \in R^2 \times [0, +\infty)$, and the function $H(z,y)$ is continuous on $[0,T]^2 \times [0,x]$. In addition, suppose that the stochastic function $\Phi(z)$ is Gaussian, has zero expectation, does not depend on $W(\Delta_1)$ and $V(\Delta_2)$ for any $\Delta_1, \Delta_2 \in [0,T]^2 \times [0,x]$, and

$$E\Phi(z)\Phi(z') = \Pi(z,z') \in L_2(R^2 \times R^2).$$

Remark 3.1 It is easy to see from (3.13) that for any $(z,x) \in \Delta$ we have

$$EX(z,x)W(\Delta) = 0,$$

$$E\int_0^x \int_{[0,T]^2} |H(z,y)X(z,y)|dzdy < \infty.$$

We investigate the following filtration problem. Suppose that we observe a random field $Y(z,y)$ on the interval $[0,T]^2 \times [0,x]$, and having these observations we aim to construct the best possible in the mean square sense estimate $\breve{X}(z,x)$, $y < x$, for the function $X(z,x)$. Denote by H^Y the linear closure (in the mean square sense) of the set of random variables $Y(\Delta)$, $\Delta \in [0,T]^2 \times [0,x]$. Since one can geometrically interpret $\breve{X}(a,x)$ as the projection of the element $\breve{X}(z,x) \in L_2$ on the subspace H^Y, then (see [22]) $\breve{X}(z,x)$ can be represented in the form

$$\breve{X}(z,x) = \int_0^x \int_{[0,T]^2} a(z,x,z',y)dY(z',y)$$

$$= \int_0^x \int_{[0,T]^2} a(z,x,z',y)H(z',y)X(z',y)dz'dy + \int_0^x \int_{[0,T]^2} a(z,x,z',y)dV(z',y),$$

$$a(z,x,\cdot,\cdot) \in L_2\left([0,T]^2 \times [0,x]\right).$$

Following the idea of innovation method [37], consider the stochastic measure $Z(\Delta)$ which satisfies for $\Delta \in [0,T]^2 \times [0,x]$ the relations

$$Z(\Delta) = \iint_\Delta dY(z,y) - \iint_\Delta H(z,y)\, \breve{X}(z,y)dzdy$$

$$= \iint_\Delta H(z,y)\widetilde{X}(z,y)dzdy + \iint_\Delta dV(z,y), \tag{3.16}$$

where $\widetilde{X} = X - \breve{X}$.

Lemma 3.4 *The random measure $Z(\Delta)$ defined by (3.14) is the Gaussian orthogonal stochastic measure on $R^2 \times [0, +\infty)$ with zero expectation and the structure function $\mu(\Delta)$.*

Proof Clearly, $Z(\Delta)$ is Gaussian. Let us prove its orthogonality. Let $\mathbf{V}_1$ be a semi-ring in R^2. If $\mathbf{V}_2$ is the class of sets of the form $[x_1, x_2) \times \Delta$, where $0 \le x_1 < x_2$, $\Delta \in \mathbf{V}_1$, then $\mathbf{V}_2$ is also semi-ring. Let $\Delta_1, \Delta_2 \in U_1$, $0 \le x_1 < x_2$, $0 \le x_3 < x_4$, $x_1 < x_4$, and denote $[x, y) = [x_1, x_2) \cap [x_3, x_4)$. Then by the definition of $Z(\Delta)$ we have

$$EZ(\Delta_1 \times [x_1,x_2)) = EZ(\Delta_2 \times [x_3,x_4)) = 0,$$

$$\begin{aligned} EZ(\Delta_1 \times [x_1,x_2))Z(\Delta_2 \times [x_3,x_4)) = EZ + EZ(\Delta_1 \times [x,y))Z(\Delta_2 \times [y,x_4)) \\ + EZ(\Delta_1 \times [x,y))Z(\Delta_2 \times [x,y))(\Delta_1 \times [x_1,x)) \\ \times Z(\Delta_2 \times [x,x_4)) \end{aligned} \tag{3.17}$$

The last summand in the right-hand side of (3.15) can be estimated as follows:

$$
\begin{aligned}
&EZ(\Delta_1 \times [x,y))Z(\Delta_2 \times [x,y)) \\
&= E\int_x^y \int_{\Delta_1} \int_x^y \int_{\Delta_2} H(z,\alpha)\widetilde{X}(z,\alpha)H(z',\beta)\widetilde{X}(z',\beta)dzd\alpha dz'd\beta \\
&\quad + \int_x^y \int_{\Delta_1} EH(z,\alpha)\widetilde{X}(z,\alpha)V(\Delta_2 \times [x,y))dzd\alpha \\
&\quad + \int_x^y \int_{\Delta_2} EH(z',\beta)\widetilde{X}(z',\beta)V(\Delta_1 \times [x,y))dz'd\beta \\
&\quad + EV(\Delta_1 \times [x,y))V(\Delta_2 \times [x,y)).
\end{aligned} \tag{3.18}
$$

For the integrand in the second summand in (3.16) we have

$$E\widetilde{X}(z,\alpha)V(\Delta_2 \times [x,y)) = E\widetilde{X}(z,\alpha)V(\Delta_2 \times [\alpha,y)) + E\widetilde{X}(z,\alpha)V(\Delta_2 \times [x,\alpha))$$

Since the measures $W(\Delta_1)$ and $V(\Delta_2)$ are orthogonal, then $E\widetilde{X}(z,\alpha)V(\Delta_2 \times [\alpha,y)) = 0$ and $E\widetilde{X}(z',\beta)V(\Delta_1 \times [\beta,y)) = 0$.

Therefore, since for any $\Delta \in R^2$ and $0 \le \tau \le \alpha$ the stochastic fields $\widetilde{X}(z,\alpha)$ and $Y(\Delta \times [0,\tau))$ are orthogonal, one can transform the first three terms in (3.16) as

$$
\begin{aligned}
&\int_x^y \int_{\Delta_1} E\left\{ H(z,\alpha)\widetilde{X}(z,\alpha)\left(\int_x^\alpha \int_{\Delta_2} H(z',\beta)\widetilde{X}(z',\beta)dz'd\beta + V(\Delta_2 \times [x,\alpha)) \right) \right\} dzd\alpha \\
&\quad + \int_x^y \int_{\Delta_2} E\left\{ H(z,\alpha)\widetilde{X}(z,\alpha)\left(\int_\alpha^y \int_{\Delta_1} H(z',\beta)\widetilde{X}(z',\beta)dz'd\beta + V(\Delta_1 \times [\alpha,y)) \right) \right\} dzd\alpha \\
&= \int_x^y \int_{\Delta_1} EZ(\Delta_2 \times [x,\alpha))H(z,\alpha)\widetilde{X}(z,\alpha)dzd\alpha \\
&\quad + \int_x^y \int_{\Delta_2} EZ(\Delta_1 \times [x,\alpha))H(z,\alpha)\widetilde{X}(z,\alpha)dzd\alpha = 0.
\end{aligned}
$$

Hence,

$$EZ(\Delta_1 \times [x,y))Z(\Delta_2 \times [x,y)) = \mu((\Delta_1 \cap \Delta_2) \times [x,y)).$$

Similarly, it is easy to show that the first and the second terms in (3.15) are equal to zero. Therefore,

$$EZ(\Delta_1 \times [x_1,x_2))Z(\Delta_2 \times [x_3,x_4)) = \mu(\Delta_1 \times [x_1,x_2) \cap \Delta_2 \times [x_3,x_4)),$$

which in turn implies that the random function $Z(\Delta)$ possesses the orthogonality property. □

From the definition of the estimate $\breve{X}(z,x)$ we have

$$Z(\Delta) = \iint\limits_{\Delta} dY(z,y) - \iint\limits_{\Delta} H(z,y)a(z,x,z',\tau))dY(z',\tau)dzdy. \tag{3.19}$$

If the eigenvalues of the kernel $EH(z,x)X(z,x)H(z',\tau)X(z',\tau)$ are nonnegative, then (see [64]) the function $Y(z,y)$ can be written as

$$Y(\Delta) = \iint\limits_{\Delta} dZ(z,y) + \iint\limits_{\Delta} \int\limits_0^x \int\limits_{[0,T]^2} EH(z,y)a(z,x,z',\tau))dZ(z',\tau)dzdy, \tag{3.20}$$

which implies that the minimal σ-algebras with respect to which $Z(\Delta)$ and $Y(\Delta)$, $\Delta \in D$, are measurable, coincide. Therefore, we don't lose the information obtained while observing the stochastic field $X(z,x)$, if we substitute the function Y with Z. Thus, we can look for the estimate in the form

$$\breve{X}(z,x) = \int\limits_0^x \int\limits_{[0,T]^2} a(z,x,z',y)dZ(z',y).$$

Taking into account that $\widetilde{X}(z,x)$ and $Z([0,T]^2 \times [0,\tau])$ are orthogonal for any z, x and τ, $0 \le \tau \le x$, we obtain

$$\begin{aligned} EX(z,x)Z\left([0,T]^2 \times [0,\tau]\right) &= E\left\{\int\limits_0^x \int\limits_{[0,T]^2} a(z,x,z',y)dZ(z',y) \int\limits_0^\tau \int\limits_{[0,T]^2} dZ(z',y)\right\} \\ &= \int\limits_0^x \int\limits_{[0,T]^2} a(z,x,z',y)dz'dy. \end{aligned} \tag{3.21}$$

Theorem 3.3 *Suppose the solution to (3.12) exists and the eigenvalues of the kernel $EH(z,x)X(z,x)H(z',\tau)X(z',\tau)$ are nonnegative. Then the linear mean square estimate $\breve{X}(z,x)$ of the random field $X(z,x)$ satisfies*

$$\breve{X}(z,x) = \int\limits_0^x L_z\, \breve{X}(z,y)dy + \int\limits_0^x \int\limits_{[0,T]^2} H(z',y)P(z',z,y)dZ(z',y) \tag{3.22}$$

with initial condition $\breve{X}(z,0) = 0$, *where*

$$P(z,z',x) := E\widetilde{X}(z,x)\widetilde{X}(z',x)$$

is a solution (in the generalized sense) to the equation

$$\frac{\partial P(z,z',x)}{\partial x} = (L_z + L_{z'})P(z,z',x) - \int\limits_{[0,T]^2} H(u,x)P(u,z,x)H(u,x)P(u,z',x)du$$

$$+ K_z^* K_{z'}^* \delta(z - z'),$$

$$P(z,z',0) = \Pi(z,z'). \tag{3.23}$$

Proof Consider the expression in the left-hand side of (3.23) and substitute Z with (3.18). We get

$$EX(z,x)Z\left([0,T]^2 \times [0,\tau]\right) = E\int\limits_0^\tau \int\limits_{[0,T]^2} X(z,x)H(z',y)\widetilde{X}(z',y)dz'dy$$

$$+ E\int\limits_0^\tau \int\limits_{[0,T]^2} X(z,x)dV(z',y).$$

Using (3.15), Remark 3.2 and Fubini's theorem, we get

$$E\int\limits_0^\tau \int\limits_{[0,T]^2} X(z,x)H(z',y)\widetilde{X}(z',y)dz'dy$$

$$= E\int\limits_0^\tau \int\limits_{[0,T]^2} \int\limits_{R^2} G(z,x,u,y)H(u,y)P(u,z',y)dudz'dy.$$

Since $\Phi(z)$ does not depend on V, and the measures V and W are orthogonal, then

$$E\left\{\int\limits_{R^2} G(z,x,z',0)\Phi(z')dz' + \int\limits_0^x \int\limits_{R^2} K_{z'}^* G(z,x,z',\tau)dW(z',\tau)\right\}V\left([0,T]^2 \times [0,\tau]\right) = 0.$$

Setting $a(z,x,z',y) := \int_{R^2} G(z,x,u,y)H(u,y)P(u,z',y)du$, we automatically get (3.21), and the estimate $\breve{X}(z,x)$ takes the form

$$\breve{X}(z,x) = \int_0^x \int_{[0,T]^2} \int_{R^2} G(z,x,u,y)H(u,y)P(u,z',y)dZ(z',y)du$$
$$= \int_0^x \int_{[0,T]^2} H(z',y)P(z',z,y)dZ(z',y) \tag{3.24}$$
$$+ \int_0^x L_z \int_0^\tau \int_{[0,T]^2} \int_{R^2} G(z,\tau,u,y)H(u,y)P(u,z',y)du\,dZ(z',y)d\tau.$$

By continuity of H and P, and the smoothness of the fundamental solution, one can easily show that the differential operator can be pulled out from the integral with respect to the stochastic measure Z, and thus the last equality is equivalent to (3.22).

Put now $\Pi(z,z',x) := EX(z,x)X(z',x)$. Then

$$\Pi(z,z',x) = \int_{R^2}\int_{R^2} G(z,x,u,0)G(z',x,v,0)\Pi(v,u,0)dvdu$$
$$+ \int_0^x \int_{R^2} K_u^* G(z,x,u,0)K_u^* G(z,x,v,0)dudv.$$

Therefore, the function $\Pi(z,z',x)$ is the solution (in generalized sense) to the equation

$$\frac{\partial \Pi(z,z',x)}{\partial x} = (L_z + L_{z'})\Pi(z,z',x) + K_z^* K_{z'}^* \delta(z-z'), \Pi(z,z',0) = \Pi(z,z'). \tag{3.25}$$

Letting $Q(z,z',x) := E\,\breve{X}(z,x)\,\breve{X}(z',x)$ and using (3.22), we get

$$\frac{\partial Q(z,z',x)}{\partial x} = (L_z + L_{z'})Q(z,z',x)$$
$$+ \int_{[0,T]^2} H(u,x)P(u,z,x)H(u,x)P(u,z',x)du, Q(z,z',0) = 0. \tag{3.26}$$

Since $\breve{X}$ and $\widetilde{X}$ are orthogonal for any z, z', we get

$$\Pi(z,z',x) = EX(z,x)X(z',x) = E\,\breve{X}(z,x)\,\breve{X}(z',x) + E\widetilde{X}(z,x)\widetilde{X}(z',x)$$
$$= Q(z,z',x) + P(z,z',x).$$

Thus, it follows from (3.25) and (3.26) that the function $P(z,z',x)$ satisfies (3.23). □

Let us briefly stop on the prediction problem. Suppose that we need to estimate a random field $u(z,x)$ in the point z, basing on the observations of the random field $\varsigma(z,x)$ on $[0,z_1] \times \Omega$, where $z_1 < z$. Then the integral equation for optimal prediction takes the form

$$Eu(z,x)\ \dot{\varsigma}(z',y) = \int\limits_{[0,z_1]} a(z,x,\sigma,\nu)E\ \dot{\varsigma}(\sigma,\nu)\ \dot{\varsigma}(z',y)d\sigma d\nu.$$

Differentiating both sides with respect to x we get

$$\frac{\partial}{\partial x}Eu(z,x)\ \dot{\varsigma}(z',y) = E\Big[L_z u(z,x)\ \dot{\varsigma}(z',y) + K_z\ \dot{W}(z,x)\ \dot{\varsigma}(z',y)\Big],$$

$$\frac{\partial}{\partial x}\int\limits_{[0,z_1]} a(z,x,\sigma,\nu)E\ \dot{\varsigma}(\sigma,\nu)\ \dot{\varsigma}(z',y)d\sigma d\nu = \int\limits_{[0,z_1]}\int\limits_{\Omega}\frac{\partial a(z,x,\sigma,\nu)}{\partial x}E\ \dot{\varsigma}(\sigma,\nu)\ \dot{\varsigma}(z',y)d\sigma d\nu.$$

Hence,

$$L_z a(z,x,\sigma,\nu) - \frac{\partial a(z,x,\sigma,\nu)}{\partial x} = 0, \sigma \leq z_1, \nu \in \Omega.$$

One can prove also the converse statement. It is easy to see from the last equality that the optimal estimate $\breve{u}(z,x)$ of the field $u(z,x)$ satisfies

$$\frac{\partial\,\breve{u}(z,x)}{\partial x} = \int\limits_{[0,z_1]}\int\limits_{\Omega}\frac{\partial a(z,x,\sigma,\nu)}{\partial x}\ \dot{\varsigma}(\sigma,\nu)d\sigma d\nu = \int\limits_{[0,z_1]}\int\limits_{\Omega} L_z a(z,x,\sigma,\nu)\ \dot{\varsigma}(\sigma,\nu)d\sigma d\nu$$
$$= L_z\ \breve{u}(z,x).$$

Thus, we have $\dfrac{\partial\ \breve{u}\ (z,x)}{\partial x} = L_z\ \breve{u}(z,x)$, with the initial condition defined by the estimate of the field $\breve{u}(z,x)$ in the point $z = z_1$. In other words, to determine the initial condition we need to solve the optimal filter equation.

Remark 3.2 One can obtain the last relation without the assumption that $\dot{V}(z,x)$ is the white noise. However, the optimal filter equation holds true only in the case when $\dot{V}(z,x)$ is the white noise.

3.4 Filtration Problem for Stochastic Fields Described by Parabolic Equations with Given Boundary Conditions

In this section we consider an approach for solving filtration problems for stochastic parabolic equations, which is a bit different from those treated in Sects. 3.2 and 3.3. The approach we are going to describe was proposed in [69].

Assume that we have a random field $u(x,t)$ generated by a system described by the equation

$$\frac{\partial u}{\partial t} = a(x,t)\frac{\partial^2 u}{\partial x^2} + b(x,t)u + W(x,t), \tag{3.27}$$

with the boundary condition $u(0,t) = 0$ and the initial condition $u(x,0) = f(x)$, where $x \in R_+$, $t \in R_+$.

Suppose that the stochastic field $u(x,t)$ is observed with an error, i.e., we observe another field $Y(x,t)$ related to $u(x,t)$ by

$$Y(x,t) = c(x,t)u(x,t) + V(x,t). \tag{3.28}$$

Having the observed values of $Y(x,t)$ we need to construct the unbiased effective estimate $\hat{u}\,(x,t)$ of the $u(x,t)$.

We make certain assumptions on the functions involved in (3.25) and (3.26). Suppose that the functions $a(x,t)$, $b(x,t)$, and $c(x,t)$ are deterministic, $a(x,t) \neq 0$, $c(x,t) \neq 0$, and the random fields $W(x,t)$, $V(x,t)$, and $f(x)$ have the following correlation properties:

$$EW(x,t) = EV(x,t) = 0,$$

$$Ef(x) = F(x),$$

$$E[W(x,t)W(\xi,\tau)] = M(x,\xi,t)\delta(t-\tau),$$

$$E[V(x,t)V(\xi,\tau)] = N(x,\xi,t)\delta(t-\tau),$$

$$E[f(x)f(\xi)] < \infty,$$

$$E[W(x,t)V(\xi,\tau)] = E[W(x,t)f(\xi)] = E[V(x,t)f(\xi)] = 0,$$

where $\delta(t) = \begin{cases} 1, t = 0 \\ 0, t \neq 0 \end{cases}$.

The deterministic functions $F(x)$, $M(x,\xi,t)$ and $N(x,\xi,t)$ are continuously differentiable in all variables on the whole domain, $M(x,\xi,t) \geq 0$, $N(x,\xi,t) > 0$.

Following Kalman's idea, we construct the estimate $\hat{u}\,(x,t)$ of the random field $u(x,t)$ as the solution to the problem

$$\frac{\partial \hat{u}}{\partial t} = a(x,t)\frac{\partial^2 \hat{u}}{\partial x^2} + \beta(x,t)\hat{u} + \gamma(x,t)Y(x,t), \hat{u}\,|_{x=0} = 0, \hat{u}\,|_{t=0} = f(x). \quad (3.29)$$

For arbitrary deterministic functions $\beta(x,t)$ and $\gamma(x,t)$, the system (3.29) describes all possible estimates of the field $u(x,t)$ based on the observations of $Y(x,t)$. On the other hand, one can try to find some functions $\beta(x,t)$ and $\gamma(x,t)$, for which the estimation error

$$\varepsilon(x,t) = u(x,t) - \hat{u}\,(x,t)$$

is unbiased, i.e.,

$$E\varepsilon(x,t) = 0$$

and is effective, which means that

$$E\varepsilon^2(x,t) = \min.$$

Let us first investigate how the functions $\beta(x,t)$ and $\gamma(x,t)$ are related, by calculating the estimation error. We have

$$\frac{\partial \varepsilon}{\partial t} = a(x,t)\frac{\partial^2 \varepsilon}{\partial x^2} + [b(x,t) - \beta(x,t) - \gamma(x,t)c(x,t)]u + \beta(x,t)\varepsilon + W(x,t) \\ - \gamma(x,t)V(x,t),$$

and taking the expectation from both sides we get

$$\frac{\partial E\varepsilon}{\partial t} = a(x,t)\frac{\partial^2 E\varepsilon}{\partial x^2} + [b(x,t) - \beta(x,t) - \gamma(x,t)c(x,t)]Eu(x,t) + \beta(x,t)E\varepsilon,$$

from where we see that the condition of unbiaseness is equivalent to

$$[b(x,t) - \beta(x,t) - \gamma(x,t)c(x,t)]Eu(x,t) = 0.$$

Thus,

$$\beta(x,t) = b(x,t) - \gamma(x,t)c(x,t) \quad (3.30)$$

is the sufficient condition for estimate to be unbiased. In the case when the solution of the boundary problem

$$\frac{\partial Eu}{\partial t} = a(x,t)\frac{\partial^2 Eu}{\partial x^2} + b(x,t)Eu,$$

$$\hat{u}\,(0,t) = 0, \hat{u}\,(x,0) = F(x),$$

is never equal to zero, condition (3.30) is also the necessary one. Taking into account (3.28) we obtain

$$\frac{\partial \varepsilon}{\partial t} = a(x,t)\frac{\partial^2 \varepsilon}{\partial x^2} + [b(x,t) - \gamma(x,t)c(x,t)]\varepsilon + W(x,t) - \gamma(x,t)V(x,t), \qquad (3.31)$$

with the boundary condition $\varepsilon(0,t) = 0$, and the initial condition $\varepsilon(x,0) = 0$.

The problem (3.31) is reversible. Denote by $G(x,t,\xi,\tau)$ the Green function for boundary problem

$$\frac{\partial \varepsilon}{\partial t} = a(x,t)\frac{\partial^2 \varepsilon}{\partial x^2}, \varepsilon(0,t) = 0.$$

Then (3.31) is equivalent to integral equation

$$\varepsilon(x,t) = \int_0^\infty \int_0^t G(x,t,\eta,\theta)([b(\eta,\theta) - \gamma(\eta,\theta)c(\eta,\theta)]\varepsilon(\eta,\theta) \\ + W(\eta,\theta) - \gamma(\eta,\theta)V(\eta,\theta))d\eta d\theta.$$

Let us introduce some notation for correlation functions:

$$R_{\varepsilon\varepsilon}(x,t,\xi,\tau) = E[\varepsilon(x,t)\varepsilon(\xi,\tau)],$$

$$R_{\varepsilon w}(x,t,\xi,\tau) = E[\varepsilon(x,t)W(\xi,\tau)],$$

$$R_{\varepsilon v}(x,t,\xi,\tau) = E[\varepsilon(x,t)V(\xi,\tau)].$$

Multiplying the last equation by $\varepsilon(\xi,\tau)$ and taking the expectation, we derive

$$R_{\varepsilon\varepsilon}(x,t,\xi,\tau) = \int_0^\infty \int_0^t G(x,t,\eta,\theta)([b(\eta,\theta) - \gamma(\eta,\theta)c(\eta,\theta)]R_{\varepsilon\varepsilon}(\eta,\theta,\xi,\tau))d\eta d\theta \\ + \int_0^\infty \int_0^t G(x,t,\eta,\theta)[R_{\varepsilon w}(\xi,\tau,\eta,\theta) - \gamma(\eta,\theta)R_{\varepsilon v}(\xi,\tau,\eta,\theta)]d\eta d\theta.$$

Similarly, it is easy to see that

$$R_{\varepsilon w}(x,t,\xi,\tau) = \int_0^\infty \int_0^t G(x,t,\eta,\theta)[b(\eta,\theta) - \gamma(\eta,\theta)c(\eta,\theta)]R_{\varepsilon w}(\eta,\theta,\xi,\tau)d\eta d\theta \\ + \int_0^\infty G(x,t,\eta,\tau)M(\xi,\eta,\tau)d\eta$$

and

$$R_{\varepsilon v}v(x,t,\xi,\tau) = \int_0^\infty \int_0^t G(x,t,\eta,\theta)[b(\eta,\theta) - \gamma(\eta,\theta)c(\eta,\theta)]R_{\varepsilon v}(\eta,\theta,\xi,\tau)d\eta d\theta$$
$$- \int_0^\infty G(x,t,\eta,\tau)\gamma(\eta,\tau)N(\xi,\eta,\tau)d\eta.$$

The three latter integral equations relate the unknown functions $R_{\varepsilon\varepsilon}(x,t,\xi,\tau)$, $R_{\varepsilon w}(x,t,\xi,\tau)$, $R_{\varepsilon v}(x,t,\xi,\tau)$, and $\gamma(x,t)$. The fourth equation can be derived from the effectiveness condition by applying the orthogonal projections theorem, see [1].

Let H be a Hilbert space consisting of random fields $u(x,t)$ with $Eu^2(x,t) < \infty$, with the scalar product $(u(x,t), v(x,t)) := Eu(x,t)v(x,t)$. Since we have the mean square metric on H, we can conclude that the estimate $\hat{u}\,(x,t)$ is effective if and only if the estimation error is orthogonal to any estimate $Z(x,t)$, i.e., $(u - \hat{u}\,, Z) = 0$.

For any observed values $Y(x,t)$ any linear estimate of the stochastic field $u(x,t)$ takes the form

$$Z(x,t) = \int_0^\infty \int_0^t P(x,t,\xi,\tau)Y(\xi,\tau)d\xi d\tau + \int_0^\infty Q(x,\xi)f(\xi)d\xi,$$

where $P(x,t,\xi,\tau)$ and $Q(x,\xi)$ are arbitrary functions. Therefore,

$$(u - \hat{u}\,, Z) = \int_0^\infty \int_0^t P(x,t,\xi,\tau)E[\varepsilon(x,t)Y(\xi,\tau)]d\xi d\tau$$
$$+ \int_0^\infty Q(x,\xi)E[\varepsilon(x,t)f(\xi)]d\xi = 0.$$

Since the functions P and Q are arbitrary, we have

$$E[\varepsilon(x,t)Y(\xi,\tau)] = 0, \qquad (3.32)$$

$$E[\varepsilon(x,t)f(\xi)] = 0.$$

Formulas (3.32) are the analogues to the Wiener-Hopf equation, which appears in filtration problems for random processes. Taking into account the representations for $Y(x,t)$ and $\varepsilon(x,t)$, one can rewrite (3.32) as

$$E([u(x,t) - \hat{u}\,(x,t)][c(\xi,\tau)u(\xi,\tau) + V(\xi,\tau)]) = 0. \qquad (3.33)$$

It is easy to see that

$$E[u(x,t)V(\xi,\tau)] = 0.$$

Indeed, it follows from (3.27) that

$$u(x,t) = \int_0^\infty \int_0^t R(x,t,\xi,\tau)W(\xi,\tau)d\xi d\tau + \int_0^\infty R(x,t,\xi,0)f(\xi)d\xi,$$

where $R(x,t,\xi,\tau)$ is the Green function for the problem

$$\frac{\partial u}{\partial t} = a(x,t)\frac{\partial^2 u}{\partial x^2} + b(x,t)u, u(0,t) = 0.$$

Since $E[f(x)f(\xi)] < \infty$, we get $E[u(x,t)V(\xi)] = 0$, which in turn implies that (3.31) can be transformed to

$$c(\xi,\tau)E[\varepsilon(x,t)u(\xi,\tau)] = E[\hat{u}\,(x,t)n(\xi,\tau)]. \tag{3.34}$$

Since $\hat{u}\,(x,t)$ is the solution to (3.27), it can be written in the form

$$\hat{u}\,(x,t) = \int_0^\infty \int_0^t S(x,t,\xi,\tau)\gamma(\xi,\tau)Y(\xi,\tau)d\xi d\tau + \int_0^\infty S(x,t,\xi,0)f(\xi)d\xi, \tag{3.35}$$

where $S(x,t,\xi,\tau)$ is the Green function for the problem

$$\frac{\partial \hat{u}}{\partial t} = a(x,t)\frac{\partial^2 \hat{u}}{\partial x^2} + \beta(x,t)\hat{u}\,, \hat{u}\,|_{x=0} = 0.$$

Thus,

$$c(\xi,\tau)E[\varepsilon(x,t)u(\xi,\tau)] = \int_0^\infty \int_0^t S(x,t,\eta,\theta)\gamma(\eta,\theta)E[Y(\eta,\theta)V(\xi,\tau)]d\eta d\theta$$
$$+ \int_0^\infty S(x,t,\eta,0)E[f(\eta)V(\eta,\tau)]d\eta.$$

Substituting (3.35) into (3.34) we derive

$$c(\xi,\tau)E[\varepsilon(x,t)u(\xi,\tau)] = \int_0^\infty \int_0^t S(x,t,\eta,\theta)\gamma(\eta,\theta)E[V(\eta,\theta)V(\xi,\tau)]d\eta d\theta$$
$$+ \int_0^\infty S(x,t,\eta,0)\gamma(\eta,\tau)N(\xi,\eta,\tau)d\eta$$

and, taking the limit as $\tau \to t$, $\xi \to x$, we arrive at

$$c(\xi,\tau)E[\varepsilon(x,t)u(\xi,\tau)] = \int_0^\infty S(x,t,\eta,t)\gamma(\eta,t)N(\xi,\eta,t)d\eta.$$

Since for Green function we have $S(x,t,\eta,t) = \delta(x-\eta)$ (see [68]), we get

$$c(\xi,\tau)E[\varepsilon(x,t)u(\xi,\tau)] = \gamma(x,t)N(x,x,t). \tag{3.36}$$

Next, we show that

$$E[\varepsilon(x,t)\hat{u}(x,t)] = 0. \tag{3.37}$$

Multiplying both sides of (3.35) by $\varepsilon(x,t)$ and taking the expectation, we get

$$E[\varepsilon(x,t)\hat{u}(x,t)] = \int_0^\infty\int_0^t S(x,t,\eta,\theta)\gamma(\eta,\theta)E[\varepsilon(x,t)Y(\eta,\theta)]d\eta d\theta$$
$$+\int_0^\infty S(x,t,\eta,0)E[\varepsilon(x,t)f(\eta)]d\eta = 0.$$

Further multiplying (3.37) by $c(\xi,\tau)$ and subtracting the result from (3.36), we derive

$$E\varepsilon^2(x,t) = \frac{\gamma(x,t)N(x,x,t)}{c(x,t)} = R_{\varepsilon\varepsilon}(x,t,x,t),$$

which gives

$$\gamma(x,t) = \frac{c(x,t)R_{\varepsilon\varepsilon}(x,t,x,t)}{N(x,x,t)}. \tag{3.38}$$

In such a way we proved

Theorem 3.4 *The coefficients $\gamma(x,t)$ and $\beta(x,t)$ of the optimal filter (3.27) are given by (3.36) and (3.28), in which*

$$R_{\varepsilon\varepsilon}(x,t,\xi,\tau) = \int_0^\infty\int_0^t G(x,t,\eta,\theta)\left\{\left[b(\eta,\theta) - \frac{c^2(\eta,\theta)R_{\varepsilon\varepsilon}(\eta,\theta,\eta,\theta)}{N(\eta,\eta,\theta)}\right]R_{\varepsilon\varepsilon}(\eta,\theta,\xi,\tau)\right.$$
$$\left.+R_{\varepsilon w}(\xi,\tau,\eta,\theta) - \frac{c(\eta,\theta)R_{\varepsilon\varepsilon}(\eta,\theta,\eta,\theta)R_{\varepsilon v}(\eta,\theta,\eta,\theta)}{N(\eta,\eta,\theta)}\right\}d\eta d\theta,$$

$$R_{\varepsilon w}(x,t,\xi,\tau) = \int_0^\infty \int_0^t G(x,t,\eta,\theta)\left\{\left[b(\eta,\theta) - \frac{c^2(\eta,\theta)R_{\varepsilon\varepsilon}(\eta,\theta,\eta,\theta)}{N(\eta,\eta,\theta)}\right]R_{\varepsilon w}(\eta,\theta,\xi,\tau)d\eta d\theta\right.$$
$$+ \int_0^\infty G(x,t,\eta,\tau)M(\xi,\eta,\tau)d\eta,$$

$$R_{\varepsilon v}(x,t,\xi,\tau) = \int_0^\infty \int_0^t G(x,t,\eta,\theta)\left\{\left[b(\eta,\theta) - \frac{c^2(\eta,\theta)R_{\varepsilon\varepsilon}(\eta,\theta,\eta,\theta)}{N(\eta,\eta,\theta)}\right]R_{\varepsilon v}(\eta,\theta,\xi,\tau)d\eta d\theta\right.$$
$$+ \int_0^\infty G(x,t,\eta,\tau)\frac{c(\eta,\tau)R_{\varepsilon\varepsilon}(\eta,\tau,\eta,\tau)}{N(\eta,\eta,\tau)}N(\xi,\eta,\tau)d\eta,$$

where $G(x,t,\xi,\tau)$ is the Green function.

Remark 3.3 It is quite natural that the system under consideration is nonlinear, because in case of filtration of random processes the filter coefficients are defined from the nonlinear Riccati equation.

3.5 Duality of Filtration and Control Problems

In this section we discuss the duality of filtration and control problems for simplest stochastic parabolic equations. Namely, we prove the duality of filtration and control problems for some deterministic system analogous to those described in pioneer papers by Kalman and Bucy by classical stochastic differential equations of one variable. We follow the presentation of results obtained in [69].

Consider the random field $u(x,t)$ which satisfies the stochastic parabolic equation

$$\frac{\partial u}{\partial t} = A(x,t)\frac{\partial^2 u}{\partial x^2} + B(x,t)u + \dot{W}(x,t) \tag{3.39}$$

with boundary conditions

$$u(a,t) = u(b,t) = 0$$

and initial condition

$$u(x,t_0) = f(x).$$

We assume that $x \in [a,b]$, $t \in [t_0, \infty)$, the fields $A(x,t)$ and $B(x,t)$ are some given deterministic functions, $A(x,t) \neq 0$ for all x and t, and $A(x,t)$ is twice differentiable in x.

Assume that $W(x,t)$ and $f(x)$ are the random fields with the following correlation properties:

$$Ef(x) = F(x),$$
$$E([f(x) - F(x)][f(\xi) - F(\xi)]) = \varphi(x)\delta(x - \xi),$$
$$E\dot{W}(x,t) = 0,$$
$$E\left[\dot{W}(x,t)\ \dot{W}(\xi,\tau)\right] = \mu(x,t)\delta(x - \xi)\delta(t - \tau),$$
$$E\dot{W}(x,t)f(\xi) = 0.$$

Here $F(x)$, $\varphi(x)$, $M(x,t)$ are the deterministic functions, and $\varphi(x) \geq 0$, $\mu(x,t) \geq 0$ on the whole domain.

We observe the values of some stochastic field $Y(x,t)$, related to $u(x,t)$ by

$$Y(x,t) = c(x,t)u(x,t) + \dot{V}(x,t),$$

where $c(x,t)$ is deterministic, and $\dot{V}(x,t)$ is a stochastic field with

$$E\ \dot{V}(x,t) = 0,$$
$$E\left[\dot{V}(x,t)\ \dot{V}(\xi,\tau)\right] = \nu(x,t)\delta(x - \xi)\delta(t - \tau),$$
$$E\left[\dot{V}(x,t)f(\xi)\right] = E\left[\dot{V}(x,t)\ \dot{W}(\xi,\tau)\right] = 0,$$

where $\nu(x,t)$ is strictly positive, and $\nu(x,t)$ is continuously differentiable.

The filtration problem can be described as follows. On the input of the system one has a signal with white noise $\dot{W}$, and on the other hand there is the outcome signal with an additional noise $\dot{V}$ with uncorrelated increments. We need to construct the optimal estimate $\hat{u}$ of the true signal u, given the values of the observed field Y. Consider this problem in detail.

Take some deterministic function $\alpha(x)$ defined on the interval $[a, b]$, and consider the functional

$$J(\alpha) = \int_a^b u(x,t_1)\alpha(x)dx,$$

where t_1 is the fixed moment of time, $t_1 > t_0$, and $u(x,t)$ is the solution to (3.34). Given the stochastic field $Y(x,t)$ which is observed during the time period $[t_0,t_1]$, we need to construct an unbiased and effective estimate $\hat{J}(\alpha)$ of $J(\alpha)$, i.e.,

$$E\left[J(\alpha) - \hat{J}(\alpha)\right] = 0 \tag{3.40}$$

$$E\left[\left(J(\alpha) - \hat{J}(\alpha)\right)^2\right] = \min. \tag{3.41}$$

Recall that the most general form of a linear estimate is

$$\hat{J}(\alpha) = \int_a^b \hat{u}(x,t_1)\alpha(x)dx = \int_{t_0}^{t_1}\int_a^b Y(x,t)\gamma(x,t)dxdt + \int_a^b \beta(x)F(x)dx,$$

Taking into account the representation of $Y(x,t)$, we can rewrite it as

$$\hat{J}(\alpha) = \int_{t_0}^{t_1}\int_a^b \left[c(x,t)u(x,t) + \dot{V}(x,t)\right]\gamma(x,t)dxdt + \int_a^b \beta(x)F(x)dx, \qquad (3.42)$$

where the integral is understood in the same sense as in Sects. 3.3 and 3.4, and the deterministic functions $\gamma(x,t)$ and $\beta(x)$ need to be defined in such a way that conditions (3.40) and (3.41) hold true.

Following the Kalman idea, we investigate the boundary value problem with deterministic coefficients:

$$\frac{\partial Z}{\partial t} = -\frac{\partial^2}{\partial x^2}[A(x,t)Z] - B(x,t)Z + c(x,t)\gamma(x,t), \qquad (3.43)$$

$$Z(a,t) = Z(b,t) = 0, Z(x,t_0) = \alpha(x).$$

Then the functional $J(\alpha)$ can be represented in the form

$$\int_a^b u(x,t_1)\alpha(x)dx = \int_a^b Z(x,t_1)u(x,t_1)dx$$

$$= \int_a^b Z(x,t_0)u(x,t_0)dx + \int_{t_0}^{t_1} d\left[\int_a^b Z(x,t)u(x,t)dx\right], \qquad (3.44)$$

and the last differential can be calculated as below:

$$d\left[\int_a^b Z(x,t)u(x,t)dx\right] = \int_a^b \frac{\partial Z}{\partial t}udxdt + \int_a^b Z\frac{\partial u}{\partial t}dxdt$$

$$= \int_a^b \left\{-\frac{\partial^2}{\partial x^2}[A(x,t)Z] - B(x,t)Z + c(x,t)\gamma(x,t)\right\}u(x,t)dxdt$$

$$+ \int_a^b Z(x,t)\left[A(x,t)\frac{\partial^2 u}{\partial x^2} + B(x,t)u + \dot{W}(x,t)\right]dxdt$$

$$= \int_a^b \left\{A(x,t)z(x,t)\frac{\partial^2 u}{\partial x^2} - \frac{\partial^2}{\partial x^2}[A(x,t)z]u(x,t)\ \dot{W}(x,t)Z(x,t)\right.$$

$$\left. + c(x,t)\gamma(x,t)\right\}dxdt.$$

Taking into account (3.39), (3.43), and applying integration by parts we obtain

$$\int_a^b \frac{\partial^2}{\partial x^2}[A(x,t)Z]u(x,t)dx = -\int_a^b \frac{\partial}{\partial x}[A(x,t)Z]\frac{\partial u}{\partial x}dx = \int_a^b A(x,t)Z\frac{\partial^2 u}{\partial x^2}dx.$$

Therefore,

$$d\left[\int_a^b Z(x,t)u(x,t)dx\right] = \int_a^b \dot{W}(x,t)Z(x,t)dxdt + \int_a^b c(x,t)\gamma(x,t)dxdt,$$

which together with (3.42) gives

$$\int_a^b u(x,t_1)\alpha(x)dx = \int_a^b Z(x,t_0)u(x,t_0)dx + \int_{t_0}^{t_1}\int_a^b \left[\dot{W}(x,t)Z(x,t) + c(x,t)u(x,t)\gamma(x,t)\right]dxdt. \quad (3.45)$$

Comparing (3.44) and (3.45) we derive

$$\int_a^b \alpha(x)[u(x,t_1) - \hat{u}(x,t_1)]dx = \int_a^b [Z(x,t_0)f(x) - \beta(x)F(x)]dx + \int_{t_0}^{t_1}\int_a^b \left[\dot{W}(x,t)Z(x,t) - \dot{V}(x,t)\gamma(x,t)\right]dxdt. \quad (3.46)$$

Taking the expectation from both parts of the last equality, we get

$$\int_a^b \alpha(x)E[u(x,t_1) - \hat{u}(x,t_1)]dx = \int_a^b [Z(x,t_0) - \beta(x)]F(x)dx.$$

Thus, in order to have (3.40), it is enough to put $\beta(x) = Z(x,t_0)$.

Remark 3.4 If we know that the function $F(x)$ is only positive or only negative for $x \in [a,b]$, then the condition $\beta(x) = Z(x,t_0)$ is also the necessary one for the estimate to be unbiased.

Taking the square of (3.44) and calculating the expectations, we derive

$$E\left\{\int_a^b \alpha(x)[u(x,t_1)-\hat{u}(x,t_1)]dx\right\}^2$$
$$=\int_a^b\int_a^b Z(x,t_0)Z(\xi,t_0)E([f(\xi)-F(\xi)][f(x)-F(x)])d\xi dx$$
$$+\int_{t_0}^{t_1}\int_{t_0}^{t_1}\int_a^b\int_a^b\left[Z(x,t)Z(\xi,\tau)E\left[\dot{W}(x,t)\ \dot{W}(\xi,\tau)\right]+\gamma(x,t)\gamma(\xi,\tau)E\left[\dot{V}(x,t)\ \dot{V}(\xi,\tau)\right]\right]d\xi d\tau dx dt$$
$$=\int_a^b Z^2(x,t_0)\Phi(x)dx+\int_{t_0}^{t_1}\int_a^b\left[Z^2(x,t)\mu(x,t)+\gamma^2(x,t)\nu(x,t)\right]dxdt.$$

Therefore, in order to show (3.41), one needs to find such a control $\gamma(x,t)$ of the system (3.42) which minimizes the criterion

$$I(\gamma)=\int_a^b Z^2(x,t_0)\phi(x)dx+\int_{t_0}^{t_1}\int_a^b\left[Z^2(x,t)\mu(x,t)+\gamma^2(x,t)\nu(x,t)\right]dxdt. \tag{3.47}$$

If the optimal control $\gamma(x,t)$ is found, it is easy to find the corresponding solution $Z(x,t)$ to the problem (3.43), and, thus, to find the function $\beta(x)$. We have proved

Theorem 3.5 *Consider for criterion $J(\alpha)$ the estimate $\hat{J}(\alpha)$ of the form (3.40). If the function $\gamma(x,t)$ provides the optimal control for the problem (3.39) with criterion (3.43), and $\beta(x) = Z(x,t_0)$, then the estimate $\hat{J}(\alpha)$ is effective and unbiased.*

Remark 3.5 If the function $F(x)$ does not have zeroes and does not change the sign on the interval $[a, b]$, then the conditions of Theorem 3.5 are also necessary.

Remark 3.6 To obtain an estimate of the field in the given point (i.e. to calculate the value $\hat{u}(x_1,t_1)$) it is sufficient to take $\alpha(x) = \delta(x_1 - x)$, $x_1 \in [a,b]$.

We solve the control problem (3.39) with criterion (3.43) by applying the standard technique of calculus of variations. Let $G(x,t,\xi,\tau)$ be the Green function for the boundary value problem

$$\frac{\partial Z}{\partial t}=-\frac{\partial^2}{\partial x^2}[A(x,t)Z]-B(x,t)Z, Z(a,t)=Z(b,t)=0.$$

Then the solution to (3.41) can be written in the form

$$Z(x,t)=\int_{t_0}^{t_1}\int_a^b H(x,t,\xi,\tau)\gamma(\xi,\tau)d\xi d\tau+\int_a^b G(x,t,\xi,t_1)\alpha(\xi)d\xi,$$

where $H(x,t,\xi,\tau) = c(\xi,\tau)G(x,t,\xi,\tau)$. Substituting it into (3.45), we derive

$$I[\gamma] = \int_a^b\int_{t_0}^{t_1}\int_a^b\int_{t_0}^{t_1}\int_a^b H(x,t_0,\xi,\tau)H(x,t_0,\eta,\theta)\varphi(x)\gamma(\xi,\tau)\gamma(\eta,\theta)d\xi d\tau d\eta d\theta dx$$
$$+\int_a^b\int_a^b\int_a^b G(x,t_0,\xi,t_1)G(x,t_0,\eta,t_1)\varphi(x)\alpha(\xi)\alpha(\eta)d\xi d\eta dx$$
$$+2\int_a^b\int_a^b\int_{t_0}^{t_1}\int_a^b H(x,t_0,\xi,\tau)G(x,t_0,\eta,\tau)\varphi(x)\gamma(\xi,t)\alpha(\eta)d\xi d\tau d\eta dx$$
$$+\int_a^b\int_{t_0}^{t_1}\int_a^b\int_{t_0}^{t_1}\int_a^b\int_{t_0}^{t_1} H(x,t,\xi,\tau)H(x,t,\eta,\theta)\mu(x,t)\gamma(\xi,\tau)\gamma(\eta,\theta)d\xi d\tau d\eta d\theta dx dt$$
$$+\int_{t_0}^{t_1}\int_a^b\int_a^b\int_a^b G(x,t,\xi,t_1)G(x,t,\eta,t_1)\mu(x,t)\alpha(\xi)\alpha(\eta)d\xi d\eta dx dt$$
$$+2\int_a^b\int_a^b\int_{t_0}^{t_1}\int_a^b H(x,t,\xi,\tau)G(x,t,\eta,t_1)\mu(x,t)\gamma(\xi,t)\alpha(\eta)d\xi d\tau d\eta dx dt$$
$$+\int_{t_0}^{t_1}\int_a^b \nu(\eta,\theta)\gamma^2(\eta,\theta)d\eta d\theta.$$

In order to find the function $\gamma(x,t)$ which minimizes the functional $I[\gamma]$, consider $I[\gamma + \lambda\varepsilon]$, where λ is some numeric parameter, and $\varepsilon(x,t)$ is an arbitrary function. Changing where necessary the order of integration, we obtain

$$I[\gamma+\lambda\varepsilon] = I[\lambda] + 2\lambda\int_{t_0}^{t_1}\int_a^b d\eta d\theta\varepsilon(\eta,\theta)\left\{\int_a^b\int_{t_0}^{t_1}\int_a^b H(x,t_0,\xi,\tau)H(x,t_0,\eta,\theta)\phi(x)\gamma(\xi,\tau)d\xi d\tau dx\right.$$
$$+\int_a^b\int_a^b H(x,t_0,\eta,\theta)G(x,t_0,\xi,t_1)\phi(x)\alpha(\xi)d\xi dx$$
$$+\int_{t_0}^{\theta}\int_a^b\int_t^{t_1}\int_a^b H(x,t,\xi,\tau)H(x,t,\eta,\theta)\mu(x,t)\gamma(\xi,\tau)d\xi d\tau dx dt$$
$$\left.+\int_{t_0}^{\theta}\int_a^b\int_a^b H(x,t,\eta,\theta)G(x,t_0,\xi,\tau)\mu(x,t)\alpha(\xi)d\xi dx dt + \nu(\eta,\theta)\gamma(\eta,\theta)\right\}$$
$$+\lambda^2\int_a^b\left[\int_{t_0}^{t_1}\int_a^b H(x,t_0,\xi,\tau)\varepsilon(\xi,\tau)d\xi d\tau\right]^2\phi(x)dx$$
$$+\lambda^2\int_a^b\int_{t_0}^{t_1}\left[\int_{t_0}^{t_1}\int_a^b H(x,t,\xi,\tau)\varepsilon(\xi,\tau)d\xi d\tau\right]^2\mu(x,t)dx dt.$$

Note that the necessary condition for $\gamma(x,t)$ to be an extremum for the functional $I[\gamma]$ is

$$\frac{\partial I[\gamma+\lambda\varepsilon]}{\partial\lambda}|_{\lambda=0}=0.$$

In this case this condition is also the sufficient one for $\gamma(x,t)$ to minimize $I[\gamma]$. Denote

$$K_1(\eta,\theta,\xi,\tau)=\int_a^b H(x,t_0,\xi,\tau)H(x,t_0,\eta,\theta)\varphi(x)dx,$$

$$K_2(\eta,\theta,\xi,\tau)=\int_{t_0}^{\tau}\int_a^b H(x,t,\xi,\tau)H(x,t,\eta,\theta)\mu(x,t)dxdt,$$

$$K_3(\eta,\theta,\xi,\tau)=\int_{t_0}^{\theta}\int_a^b H(x,t,\xi,\tau)H(x,t,\eta,\theta)\mu(x,t)dxdt,$$

$$K(\eta,\theta,\xi,\tau)=\begin{cases}K_1(\eta,\theta,\xi,\tau)+K_2(\eta,\theta,\xi,\tau), t_0\leq\tau\leq\theta,\\ K_1(\eta,\theta,\xi,\tau)+K_3(\eta,\theta,\xi,\tau), \theta\leq\tau\leq t_1,\end{cases}$$

$$k(\eta,\theta)=-\int_a^b\int_a^b H(x,t_0,\eta,\theta)G(x,t_0,\xi,t_1)\phi(x)\alpha(\xi)d\xi dx$$
$$-\int_{t_0}^{\theta}\int_a^b\int_a^b H(x,t_0,\eta,\theta)G(x,t,\xi,t_1)\mu(x,t)\alpha(\xi)d\xi dxdt.$$

Using this notation we transform the condition $\frac{\partial I[\gamma+\lambda\varepsilon]}{\partial\lambda}|_{\lambda=0}=0$ into the equation

$$\gamma(\eta,\theta)\nu(\eta,\theta)+\int_{t_0}^{t_1}\int_a^b K(\eta,\theta,\xi,\tau)\gamma(\xi,\tau)d\xi d\tau=k(\eta,\theta),$$

where we used that $\varepsilon(x,t)$ is arbitrary. Since $\nu(x,t)>0$, we can divide the last equation by $\nu(x,t)$, and derive in such a way the Fredholm integral equation of the second kind:

$$\gamma(\eta,\theta)+\int_{t_0}^{t_1}\int_a^b \frac{K(\eta,\theta,\xi,\tau)}{\nu(\eta,\theta)}\gamma(\xi,\tau)d\xi d\tau=\frac{k(\eta,\theta)}{\nu(\eta,\theta)}. \tag{3.48}$$

It is known that the sufficient condition for the existence and uniqueness of the solution to (3.46) (see [18]) is given by the inequality

$$\int_{t_0}^{t_1}\int_{a}^{b}\int_{t_0}^{t_1}\int_{a}^{b}\left[\frac{K(x,t,\xi,\tau)}{\nu(x,t)}\right]^2 d\xi d\tau dx dt \le 1. \tag{3.49}$$

If inequality (3.47) holds true, the solution to (3.46) can be derived by the Picard iteration procedure.

Example 3.1 Consider the situation when $A(x,t) \equiv 1$, $B(x,t) \equiv 0$, $c(x,t) \equiv 1$, $\mu(x,t) \equiv 1$, $\varphi(x) \equiv 0$, $a = 0$, $b = 1$. Suppose also that there exists a constant N such that $\nu(x,t) \ge N > 0$ for all admissible x and t. The problem is to find the values of N that guarantee the existence and uniqueness of the solution to (3.48).

In this example the dual system (3.41) takes the form

$$\frac{\partial Z}{\partial t} = -\frac{\partial^{2Z}}{\partial x^2} + \gamma(x,t), Z(0,t) = Z(1,t) = 0, Z(x,t_1) = \alpha(x),$$

the respective Green function is

$$G(x,t,\xi,\tau) = -2\sum_{n=1}^{\infty} \sin \pi n x \sin \pi n \xi \exp\left\{(\pi n)^2 (t-\tau)\right\},$$

and since $c(x,t) \equiv 1$, we have $H(x,t,\xi,\tau) = G(x,t,\xi,\tau)$. Using last relationship, we obtain

$$K_1(x,t,\xi,\tau) = 0,$$

$$K_2(x,t,\xi,\tau) = \frac{1}{\pi^2}\sum_{n=1}^{\infty} \frac{\sin \pi n x \sin \pi n \xi}{n^2}\left\{\exp\left[(\pi n)^2(\tau - t)\right] - \exp\left[(\pi n)^2(2t_0 - \tau - t)\right]\right\},$$

$$K_3(x,t,\xi,\tau) = \frac{1}{\pi^2}\sum_{n=1}^{\infty} \frac{\sin \pi n x \sin \pi n \xi}{n^2}\left\{\exp\left[(\pi n)^2(t - \tau)\right] - \exp\left[(\pi n)^2(2t_0 - \tau - t)\right]\right\},$$

where the convergence of the is guaranteed by the restrictions $t_0 \le \tau \le t$ for the function K_2, and $t_0 \le t \le \tau$ for K_3. Further,

$$\int_{t_0}^{t_1}\int_{a}^{b}\int_{t}^{t_1}\int_{a}^{b} K^2(x,t,\xi,\tau) d\xi d\tau dx dt = \frac{t_1 - t_0}{8\pi^6}\sum_{n=1}^{\infty}\frac{2 + \exp\left[2(\pi n)^2(t_0 - t_1)\right]}{n^6}$$

$$+ \frac{3}{16\pi^8}\sum_{n=1}^{\infty}\frac{\exp\left[2(\pi n)^2(t_0 - t_1)\right] - 1}{n^8}.$$

The first series involve only positive terms, the second one—only negative, implying the estimate

$$-\frac{3}{16\pi^8}\sum_{n=1}^{\infty}\frac{1}{n^8} \leq \int_{t_0}^{t_1}\int_{a}^{b}\int_{t}^{t_1}\int_{a}^{b} K^2(x,t,\xi,\tau)d\xi d\tau dx dt \leq \frac{3(t_1-t_0)}{8\pi^6}\sum_{n=1}^{\infty}\frac{1}{n^6}.$$

Recall that

$$\sum_{n=1}^{\infty}\frac{1}{n^6} = \frac{\pi^6}{945}, \text{ and } \sum_{n=1}^{\infty}\frac{1}{n^8} = \frac{\pi^8}{9450}.$$

As $\nu(x,t) \geq N > 0$, condition (3.47) is equivalent to the system of inequalities

$$N^2 > \frac{1}{50400}, N^2 > \frac{t_1 - t_0}{2520}.$$

If N satisfies these two inequalities, then the solution to (3.48) exists and is unique.

3.6 Prediction Problem for Stochastic Fields

In this section we consider the problem of optimal in a certain sense prediction for stochastic fields. In particular, we derive the Wiener-Hopf equations and provide some examples, in which the solutions can be obtained explicitly. For the results presented below we refer to [40].

Consider the equation

$$Lu = \dot{W}(z), z = (t,s), \tag{3.50}$$

where L is an elliptic-type second-order linear differential operator, and $\dot{W}(z)$, $z \in R^2$, was defined in the Sect. 3.2.

Denote by $G(z,z_1)$ the fundamental solution to the problem

$$Lu = \delta(z), \tag{3.51}$$

where $\delta(z)$ is the δ-function concentrated at the point $z = 0$. Then the stochastic field

$$u(z) = \int_{R^2} G(z_1,z)W(dz_1) \tag{3.52}$$

is the solution to (3.50) in the generalized sense, i.e., it satisfies (3.51). Indeed,

$$\int_A L(f(z))u(z)dz = \int_A \left[L(f(z)) \int_{R^2} G(z_1, z)W(dz_1) \right] dz = \int_{R^2} f(z)W(dz)$$
$$= \int_A f(z)W(dz)$$

for any function $f(z)$ supported on a compact set A.

Let us calculate the correlation function for $u(z)$:

$$r(z_1, z_2) = E[u(z_1)u(z_2)] = E \int_{R^2} G(z_1, z)W(dz_1) \int_{R^2} G(z_2, z)W(dz_2)$$
$$= \int_{R^2}\int_{R^2} G(z_1, z')G\left(z_2, z''\right)E\left[W(dz')W\left(dz''\right)\right] = \int_{R^2} G(z_1, z)G(z, z_2)dz.$$

One can show that $r(z_1,z_2)$ satisfies the equation

$$L_{z_1}[L_{z_1}[r(z_1, z_2)]] = \delta(z_1 - z_2), \tag{3.53}$$

which means that for any compact set A we have for $f(z) \in S_2(D)$

$$\int_D L_{z_1}[L_{z_1}[f(z_1, z_2)]]r(z_1, z_2)dz_1 = \int_D \left[L_{z_1}L_{z_1}f(z_1) \int_{R^2} G(z_1, z)G(z, z_2)dz \right] dz_1$$
$$= \int_{R^2} G(z, z_2) \left[\int_D L_{z_1}L_{z_1}[f(z_1)]G(z_1, z)dz_1 \right] dz$$
$$= \int_{R^2} G(z, z_2)L[f(z)]dz = f(z_2).$$

Now we can proceed with the prediction problem. In what follows we assume that L is the second-order operator.

Suppose that we know the values of the stochastic field $u(z)$ which satisfies (3.50) in some domain D with boundary Γ. We need to estimate the values of the field in some point $z_1 = (s_1,t_1) \notin D$. Following [35], we are looking for the prediction $u(z_1)$ in the special form

$$\widehat{u}(z_1) = \int_D u_t'(z)f_{10}(z)dz + \int_D u_s'(z)f_{01}(z)dz + \int_\Gamma u_t'(z)f_1(z)dl$$
$$+ \int_\Gamma u_s'(z)f_2(z)dl. \tag{3.54}$$

Integrating by parts one can rewrite (3.52) as follows:

$$\widehat{u}(z_1) = \int_D u(z)f(z)dz + \int_\Gamma u'_t(z)f_1(z)dl + \int_\Gamma u'_s(z)f_2(z)dl + \int_\Gamma u(z)f_3(z)dl. \tag{3.55}$$

Let us find in such a class of estimates the one which minimizes $E|u(z_1) - \widehat{u}(z_1)|^2$. It can be checked that to do that we need to require that the functions $f_1(z), f_2(z), f_3(z)$ to satisfy the integral equations below:

$$\int_D r(z',z)f(z')dz' + \int_\Gamma \frac{\partial r(z',z)}{\partial t'} f_1(z')dl_1 + \int_\Gamma \frac{\partial r(z',z)}{\partial s'} f_2(z')dl_1 + \int_\Gamma r(z',z)f_3(z')dl_1 = r(z_1,z), z \in D,$$

$$\int_D \frac{\partial^2 r(z',z)}{\partial t \partial s'} f(z')dz' + \int_\Gamma \frac{\partial^2 r(z',z)}{\partial t \partial t'} f_1(z')dl_1 + \int_\Gamma \frac{\partial^2 r(z',z)}{\partial t \partial s'} f_2(z')dl_1 + \int_\Gamma \frac{\partial r(z',z)}{\partial t} f_3(z')dl_1 = \frac{\partial r(z_1,z)}{\partial t},$$

$$\int_D \frac{\partial r(z',z)}{\partial s} f(z')dt' + \int_\Gamma \frac{\partial^2 r(z',z)}{\partial s \partial t'} f_1(z')dl_1 + \int_\Gamma \frac{\partial^2 r(z',z)}{\partial s \partial s'} f_2(z')dl_1 + \int_\Gamma \frac{\partial r(z',z)}{\partial s} f_3(z')dl_1 = \frac{\partial r(z_1,z)}{\partial s},$$

$$\int_D r(z',z)f(z')dz' + \int_\Gamma \frac{\partial r(z',z)}{\partial t'} f_1(z')dl_1 = \int_\Gamma \frac{\partial r(z',z)}{\partial t'} f_2(z')dl_1 + \int_\Gamma r(z',z)f_3(z')dl_1 = r(z_1,z).$$

Applying the operator L_z^2 to the first equation we obtain $f(z) \equiv 0$. Therefore,

$$\int_\Gamma \frac{\partial r(z',z)}{\partial t'} f_1(z')dl_1 + \int_\Gamma \frac{\partial r(z',z)}{\partial s'} f_2(z')dl_1 + \int_\Gamma r(z',z)f_3(z')dl_1 = r(z_1,z),$$

$$\int_\Gamma \frac{\partial^2 r(z',z)}{\partial t \partial t'} f_1(z')dl_1 + \int_\Gamma \frac{\partial^2 r(z',z)}{\partial t \partial s'} f_2(z')dl_1 + \int_\Gamma \frac{\partial r(z',z)}{\partial t} f_3(z')dl_1 = \frac{\partial r(z_1,z)}{\partial t},$$

$$\int_{\Gamma} \frac{\partial^2 r(z',z)}{\partial s \partial t'} f_1(z')dl_1 + \int_{\Gamma} \frac{\partial^2 r(z',z)}{\partial s \partial s'} f_2(z')dl_1 + \int_{\Gamma} \frac{\partial r(z',z)}{\partial s} f_3(z')dl_1 = \frac{\partial r(z_1,z)}{\partial s}.$$

We consider in detail the case where D is the half-plane $s \le s_0$. From (3.53) we obtain

$$\widehat{u}(z_1) = \int_{-\infty}^{\infty} \int_{-\infty}^{s_0} u(z) f(z) dz + \int_{-\infty}^{\infty} u'_s(t,s_0) f_1(t,s_0) dt + \int_{-\infty}^{\infty} u(t,s_0) f_2(t,s_0) dt$$

In such a way, we derive two integral equations:

$$\int_{-\infty}^{\infty} \frac{\partial^2 r(t',s_0,t,s_0)}{\partial t \partial t'} f_1(t',s_0) dt' = \frac{\partial r(t_1,s_1,t,s_0)}{\partial t},$$

$$\int_{-\infty}^{\infty} r(t',s_0,t_1,s_0) f_2(t',s_0) dt' = r(t_1,s_1,t,s_0). \tag{3.56}$$

Suppose that the field under consideration is homogeneous. Then $r(z_1,z_2) = R(t_1 - t_2, s_1 - s_2)$, and

$$\frac{\partial r(z_1,z_2)}{\partial s_1} = \frac{\partial R(t_1-t_2,s_1-s_2)}{\partial s_1},$$

$$\frac{\partial r(z_1,z_2)}{\partial s_2} = \frac{\partial R(t_1-t_2,s_1-s_2)}{\partial s_2},$$

$$\frac{\partial^2 r(z_1,z_2)}{\partial s_1 \partial s_2} = \frac{\partial^2 R(t_1-t_2,s_1-s_2)}{\partial s_1 \partial s_2}.$$

Suppose that the functions $f_1(t,s_0), f_2(t,s_0)$ and $R(t,s_0)$ are square integrable. Then, using the Fourier transform technique, one can easily derive the solutions to (3.56).

Example 3.2. Let

$$L = \frac{\partial^2}{\partial t^2} + \frac{\partial^2}{\partial s^2} - \nu^2.$$

Applying the considerations similar to those given above we find the solution to (3.47):

$$u(z) = \frac{1}{\sqrt{2\pi}} \int_{R^2} K_0 \nu \sqrt{(t-t')^2 + (s-s')^2} W\left(dt, ds'\right),$$

where K_υ is the modified Bessel function of the third kind [2] in particular,

$$K_0(\rho) = \int_1^\infty \frac{e^{-\rho\xi}}{\sqrt{\xi^2 - 1}} d\xi, K_1(\rho) = \frac{1}{2} e^{-\pi i} \int_{-\infty}^{\infty} e^{-i\rho \sinh \xi - \xi} d\xi$$

The respective correlation function is

$$r(z_1, z_2) = \frac{\sqrt{(t_1 - t_2)^2 + (s_1 - s_2)^2}}{4\pi\nu} K_1\left(\nu\sqrt{(t_1 - t_2)^2 + (s_1 - s_2)^2}\right).$$

One can check that all required conditions are satisfied, and thus we can find the functions $f_1(t,s_0)$ and $f_2(t,s_0)$, which are the solutions to the integral equations (3.54):

$$f_1(t, s_0) = \frac{(s_1 - s_0)\nu}{\pi} \frac{K_1\left(\nu\sqrt{(t_1 - t)^2 + (s_1 - s_0)^2}\right)}{\sqrt{(t_1 - t)^2 + (s_1 - s)^2}},$$

$$f_2(t, s_0) = \frac{(s_1 - s_0)\nu^2}{\pi} \frac{K_1\left(\nu\sqrt{(t_1 - t)^2 + (s_1 - s_0)^2}\right)}{(t_1 - t)^2 + (s_1 - s_0)^2}.$$

Finally, we arrive at

$$\widehat{u}(z_1) = \int_{-\infty}^{\infty} \frac{(s_1 - s_0)^2 \nu}{\pi} \frac{K_1\left(\nu\sqrt{(t_1 - t)^2 + (s_1 - s_0)^2}\right)}{\sqrt{(t_1 - t)^2 + (s_1 - s)^2}} u_s'(t, s_0) dt$$

$$+ \int_{-\infty}^{\infty} \frac{(s_1 - s_0)^3 \nu^2}{\pi} \frac{K_1\left(\nu(t_1 - t)^2 + (s_1 - s_0)^2\right)}{(t_1 - t)^2 + (s_1 - s)^2} u(t, s_0) dt$$

Chapter 4
Control Problem for Diffusion-Type Random Fields

In this chapter we derive the conditions which guarantee the existence of optimal or ε-optimal controls for stochastic systems described by stochastic parabolic differential equation. For random processes similar problems were investigated in [26]. Control problem for some types of processes and fields was discussed also in [18]. Our references for this chapter are [13, 15, 46].

4.1 Existence of an Optimal Control

Consider the equation:

$$\xi(z) = \xi_0 + \int_{[0,z]} a(z_1, \xi, u) dz_1 + \int_{[0,z]} b(z_1, \xi) W(dz_1), z \in [0,T]^2, \tag{4.1}$$

where $a(z,x,u) : [0,T]^2 \times C[0,T]^2 \times U \to R$, $b(z, x(\cdot)) : \ [0,T]^2 \times C[0,T]^2 \to R$, are non-anticipating functions, $u = u(z)$ is a control with values in a compact metric space $(U, \mathcal{U})$ and does not depend on the future, and W is a Wiener field.

Denote by $S(C,L)$ the class of linearly bounded functions satisfying the uniform Lipschitz condition in x. Further, denote by $\boldsymbol{U}$ the class of all fields $u = u(z)$ for which there exists a strong solution to (4.1). This class is obviously nonempty: if $a(z,\cdot,u) \in S(C,L)$, $b(z,\cdot) \in S(C,L)$, with respective constants independent of u and z, then for any $\mathfrak{I}_z$-adapted field $u = u(z)$ (4.1) has a unique strong solution.

We chose the optimal control in such a way that it minimizes the cost function

$$F(u) = F_0(\xi^u(\cdot), u(\cdot)) = E_u \int_{[0,T]^2} f(z, \xi^u(z), u(z)) dz), \tag{4.2}$$

where $\xi^u(z)$ denotes the solution to (4.1), corresponding to the control $u = u(z)$ and the initial condition ξ_0. Here E_u denotes the expectation with respect to the measure,

P.S. Knopov and O.N. Deriyeva, *Estimation and Control Problems for Stochastic Partial Differential Equations*, Springer Optimization and Its Applications 83,
DOI 10.1007/978-1-4614-8286-4_4, © Springer Science+Business Media New York 2013

corresponding to the control u, and $\left(\mathcal{B}_z, z \in [0,T]^2\right)$ is the minimal σ-algebra in $C[0,T]^2$, generated by cylindrical sets D with bases in $[0,z]$, i.e. $D = \{x(\cdot) : (x(z_1), \ldots, x(z_n)) \in \mathfrak{R}\}$, where $\mathfrak{R}$ is the Borel σ-algebra on R, $0 \le z_1 < \ldots < z_n \le z$.

Now we define the optimal control as the solution to (4.1) on the set $[z,(T,T)]$.

Suppose that $f(z,x,u)$ is some jointly continuous nonnegative function defined on $(z,x,u) : [0,T]^2 \times C[0,T]^2 \times U$ and satisfying the following conditions:

(a) The function $f(z,x,u) : [0,T]^2 \times C[0,T]^2 \times U$ is bounded, $0 \le f(z,x,u) \le c$.
(b) For any (z,u) the function $f(z, x(.), u)$ is $\mathcal{B}_z$—adapted.

Definition 4.1 The value $Z_1 = \inf\limits_{u \in \mathbf{U}_1} F(u)$ is called the optimal cost (or the cost) of the control u in the class $\mathbf{U}_1$ of all measurable controls.

Definition 4.2 The value $Z_2 = \inf\limits_{u \in \mathbf{U}_2} F(u)$ is called the optimal cost (or the cost) of the control u in the class $\mathbf{U}_2$ of all controls with feedback.

Definition 4.3 If the cost function $F(u)$ attains its minimum at some $u = \eta$, then the control η is called optimal in $\mathbf{U}_1$. The control η^1 for which $F(\eta^1) - Z_1 < \varepsilon$ is called ε-optimal.

It will be shown below that the class $\mathbf{U}_\theta$ of step controls is dense in $\mathbf{U}_1$. Moreover, for any step control from $\mathbf{U}_\theta$ there exists a feedback control which is equally good or better. Therefore, class $\mathbf{U}_2$ is dense in $\mathbf{U}_1$, and $Z_1 = Z_2$. But then there arises the question: does there exist the solution to the respective stochastic differential equation, if we take $\mathbf{U}_2$ for the set of admissible controls? In this case one should understand the solution to the stochastic differential equation in the weak sense, which might exist, if the drift function contains the step controls from $\mathbf{U}_1$. The existence and uniqueness conditions for the weak solution to a stochastic differential equation are given in Chap. 2. Therefore, the control contained in the drift function in (4.1) must be such that these conditions hold true.

We need the following statement which is the generalization of the Gronwall–Bellman inequality in R^2.

Lemma 4.1 [66] *Suppose that the function $u(t, s)$ is bounded on $[0,T]^2$, and that the inequality*

$$u(t,s) \le a(t,s) + \int_0^t \int_0^s a(x,y)u(x,y)dxdy$$

holds true. Then

$$u(t,s) \le \hat{a}\,(t,s)\exp\left\{\int_0^t \int_0^s a(x,y)dxdy\right\}$$

Now we investigate the compactness problem for the family of densities $\{\rho(\zeta_{a(\cdot,u)}(T)),\ u \in U_1\}$, where $a(\cdot,u) = a(z, x(\cdot), u)$

Lemma 4.2 *Let the function $b(z, x(\cdot))$ be bounded,* $|b(z,x(\cdot))| \le c$ *and* $\left|\overline{b}^{-1}(z,x(\cdot))a(z,x(\cdot),u)\right|^2 \le c\left(1+\|x\|_z^2\right)$. *Then for some $p \ge 1$*

$$\sup_u E\exp\left\{p\zeta_{a(\cdot,u)}(T)\right\} \le K.$$

Proof It is easy to see that

$$E\exp\left\{p\zeta_{a(\tilde{n},u)}(T)\right\} \le \rho\left(\zeta_{a(\cdot,u)}(T)\right)\exp\left\{\frac{p^2-p}{2}\int_{[0,T]}^{T}\left(\overline{b}^{-1}(z,x(\cdot))a(z,x(\cdot),u)\right)^2 dz\right\}$$

$$\le \rho\left(\zeta_{a(\tilde{n},u)}(T)\right)\exp\left\{\frac{p^2-p}{2}cT^2\left(1+\|x\|_T^2\right)\right\}.$$

From Lemma 2.2 we have $E\rho(\zeta_{a(\cdot,u)}(T)) = 1$, implying that

$$x(z) = W(z) - p\int_{[0,z]} a(z',u)dz'$$

is a Wiener field with respect to the probability measure $\widetilde{P} = \rho\left(\zeta_{pa(\tilde{n},u)}(T)\right)P$. Therefore $x(z)$ is the solution to the stochastic equation

$$dx(z) = pa(z,x(\cdot),u)dz + b(z,x(\cdot))dW(z).$$

Let $\mu(z) = \int_{[0,z]} b(z, x(\cdot))dW(z)$. Then $\left\{\mu(z),\mathcal{B}_z,\widetilde{P}\right\}$ is the continuous square integrable strong martingale. We have

$$x(z) = x(0) + \int_{[0,z]} pa(z',x(\cdot),u)dz' + \mu(z),$$

from where one can derive that

$$\|x(\cdot)\|_z \le 3\left(\|x(0)\|^2 + T^2p\int_{[0,z]}|a(z',x(\cdot),u)|^2dz' + \|\mu(\cdot)\|_z\right)$$

$$\le 3\left(\|x(0)\|^2 + cT^4p^2 + \|\mu(\cdot)\|_T^2 + 3T^2p^2\int_{[0,z]}\|x(\cdot)\|_{z'}^2dz'\right).$$

Using Lemma 4.1 we obtain

$$E\exp\left\{p\zeta_{a(\cdot,u)}(T)\right\} \leq E_{\widetilde{P}}\exp\left\{\frac{p^2-p}{2}\left(A_0 + A_1\|\mu(\cdot)\|_T^2\right)\right\},$$

where A_0 and A_1 are some constants (see [26]). Since the function $\exp\{A_1\|\mu(\cdot)\|_T{}^2\}$ is the strong martingale, it follows from Theorem 1.14 that

$$\begin{aligned} E_{\widetilde{P}}\exp\left\{\frac{p^2-p}{2}A_1\|\mu(\cdot)\|_T^2\right\} &= E_{\widetilde{P}}\exp\left\{\frac{p^2-p}{2}A_1|\mu(z)|^2\right\} \\ &\leq 16E_{\widetilde{P}}\exp\left\{\frac{p^2-p}{2}A_1|\mu(T)|^2\right\}. \end{aligned}$$

Finally, by the same arguments as in ([26], Lemma 3.19), we obtain statement of the lemma. □

Corollary 4.1 *If the conditions of Lemma 4.2 hold true, then the family of densities* $\{\rho(\zeta_{a(\cdot,u)}(T)),\ u \in \mathbf{U}_1\}$ *is uniformly integrable.*

Corollary 4.2 *Under conditions of Lemma 4.2, the family* $\{\rho(\zeta_{a(\cdot,u)}(T)),\ u \in \mathbf{U}_1\}$ *is relatively weakly compact.*

Following the ideas from [26] one can show that Corollaries 4.1 and 4.2 imply the existence of the optimal control in the class $\mathbf{U}_1$.

Now we define the optimal control as the solution to (4.1) on the set $(z,T]$.

Denote

$$F(z,u) = E_u\left[\int_{(z,(T,T)]} f(z_1,\xi^u(z_1),u(z_1))dz_1)\right.$$

$$Z(z,x(\cdot)) = \inf_{u\in\mathbf{U}_1} E\int_{(z,(T,T)]} f(z_1,x(\cdot),u)dz_1 \tag{4.3}$$

For simplicity we write $Z(z) = Z(z, x(\cdot))$ where it does not cause misunderstanding.

Lemma 4.3 *The control* $u^* \in \mathbf{U}_1$ *is optimal if and only if for any* $u \in \mathbf{U}_1$ *there exists an integrable* $\mathcal{B}_z$*-consistent stochastic field* $c_u(z)$*, satisfying the conditions*

$$E_u\left\{\int_{[0,T]^2} c_u(z)dz/\mathbf{B}_0^*\right\} = J^*, \tag{4.4}$$

$$\inf_u\left(f(z,x,u) - c_u(z)\right) = f(z,x,u^*) - c_{u^*}(z) = 0, \tag{4.5}$$

where J^* *is constant.*

Proof Suppose that the control u^* is optimal. Let $J^* = F(u^*)$ and $k(u) = \frac{J^*}{F(u)}$. For every u we have $k(u) \le 1$, moreover, $k(u) = 1$ if and only if u is optimal. Put $c_u(z) = k(u)f(z,x,u)$. We obtain

$$E_u\left\{\int_{[0,T]^2} c_u(z)dz/B_0{}^*\right\} = k(u)F_0(u)^0 = J^*$$

where J^* is independent of u. Moreover,

$$f(z,x,u) - c_u(z) = (1-k(u))f(z,x,u) \ge f(z,x,u^*) - c_{u^*}(z) \equiv 0,$$

which proves the necessity.

Assume that the conditions of the lemma are fulfilled. Put $q_u(z) = \int_{[z,(T,T)]} c_u(z_1)\, dz_1$ and $Z(z,u) = E_u\{q_u(T)/B_z{}^*\}$. We obtain that $Z((0,0),u) = J^*$ is independent of u, and

$$F_z(u) - Z_z(u) = E_u\left\{\int_{[z,(T,T)]} (f(z,x,u) - c_u(z))dz/B_z{}^*\right\} \ge 0,$$

where the equality takes place for u^* and $J^* = \min_u F(u)$. □

Definition 4.4 The control u is called monotone if $(Z(z,x(\cdot),B_z,P_u)$ is a strong super-martingale. If the control u^* is optimal, then it is monotone,

Lemma 4.4 *Suppose that the control* u *is monotone. Then there exists unique* $\mathcal{B}_z$ *-adapted stochastic fields A(z) and B(z), such that*

$$\begin{aligned} &P\left\{\int_{[0,z]} |A(z')|dz' < \infty\right\} = 1, \\ &P\left\{\int_{[0,z]} B^2(z')dz'\right\} = 1 P_u\text{-a.s.} \end{aligned} \tag{4.6}$$

and

$$Z(z) = J^* + \int_{[0,z]} A(z')dz' + \int_{[0,z]} B\left(z'\right)x\left(dz'\right)P_u\text{-a.s} \tag{4.7}$$

Proof Since $(Z(z, x(\cdot))$ is a strong super-martingale, then it is easy to show that $(E_uZ(z, x(\cdot))$ is a continuous function with respect to z, and thus by Theorem 1.15 we obtain

$$Z(z) = \xi(z) + \alpha(z), \tag{4.8}$$

where $\xi(z)$ is the integrable strong martingale, $\alpha(z)$ is the $\mathcal{B}_z$-measured stochastic field, $\alpha(t,0) = \alpha(0,s) = 0$, $\alpha(z, z_1] \ge 0$, and for any $z_1 \ge z$ we have sup $E\alpha(z) < \infty$. By Theorem 1.12 there exist $\mathcal{B}_z$—adapted stochastic fields $\gamma(z)$ and $\varphi(z)$, for which (4.6) holds true, and such that

$$\alpha(z) = \gamma_0 + \int_{[0,z]} \gamma(z_1)dz_1$$

$$\xi(z) = \int_{[0,z]} \varphi(z_1)W(dz_1),$$

where $\gamma_0 = Z_0$. Since $\gamma_0 = Z_0$,

$$W(dz) = \left(\overline{b}^{-1}(z,x)\right)^2 (x(dz) - a(z,x,u)dz),$$

we obtain with μ_u-probability 1

$$Z(z) = J^* + \int_{[0,z]} A(z')dz' + \int_{[0,z]} B(z')x(dz'),$$

where $A(z) = -\gamma(z) + \overline{b}^{-1}(z,x)a(z,x,u), B(z) = \varphi(z)\overline{b}^{-1}(z,x)$. Lemma is proved.

Next we shall prove the existence of optimal control in the class $\boldsymbol{U}_1$ using ideas proposed in [26]. Using Lemma 4.4 and following ([26], Theorem 3.25) we obtain the next proposition.

Theorem 4.1 *Let optimal control exist. Then the control $u^* = u^*(z, x(\cdot))$ is optimal if and only if there exists a constant value* F* *and for any monotone control there exist some functions $\gamma^{(u)}(z)$ and $\beta^{(u)}(z)$ that*

(a) $\int_{[0,T]^2} |\beta^{(u)}(z)|dz < \infty, P_u\text{-a.s.}, E_u \int_{[0,T]^2} \beta^{(u)}(z)dx(z) = 0;$

(b) $V^{(u)}(T) = 0, \quad \text{where} \quad V^{(u)}(z) = F^* + \int_{[0,z]} \gamma^{(u)}(z)dz + \int_{[0,z]} \beta^{(u)}(z)dz;$

(c) $\inf_u \left|\gamma^{(u)}(z) + \beta^{(u)}(z)a(z,x(\cdot),u(z,x(\cdot))) + f(z,x(\cdot),u(z,x(\cdot)))\right|$
$= \left|\gamma^{(u^*)}(z) + \beta^{(u^*)}(z)a(z,x(\cdot),u^*(z,x(\cdot))) + f(z,x(\cdot),u^*(z,x(\cdot)))\right| = 0$

for almost all (z,x).

If conditions of the Theorem 4.1 hold, then F* *is an optimal cost.*

Define for any p ∈ R *and* (z,x,u) ∈ *[0,T]²* × C*[0,T]²* × U *the Hamilton function*

$$H^1(z,x,u,p) = pa(z,x,u) + f(z,x,u).$$

Let

$$H(z,x,p) := \min_{u \in U} H^1(z,x,u,p) = H^1(z,x,u_0,p). \tag{4.9}$$

Remark 4.1 Note that this minimum exists due to continuity of *a(z,x,u)*, *f(z,x,u)* and compactness U. Thus, for any *(z,x,p)* one can define the Borel function *y** *(z, x, B(z,x))* on which the minimum in (4.9) is attained.

Theorem 4.2 *Suppose that the above conditions are satisfied. Then the optimal control exists and is given by*

$$u^*(z,x) = y^*(z,x,B(z,x)) \tag{4.10}$$

Proof We show that $u^* \in$ U. Indeed, for fixed (z,x,p) the function H^1 is continuous in u, and for any fixed $u \in$ U it is measurable in (z,x,p) with respect to the σ-algebra $\mathfrak{R}_{[0,T]^2} \times B \times \mathfrak{R}_R$, where $\mathfrak{R}_{[0,T]^2}$ is the Borel σ-algebra on $[0,T]^2$, $\mathfrak{R}_R$ is the Borel σ-algebra on R. Let S be a countable everywhere dense subset of U. By the continuity of H^1 in u,

$$H(z,x,p) = \inf_{u \in S} H^1(z,x,u,p),$$

and

$$\{(z,x,p) : H(z,x,p) < a\} = \bigcup_{u \in S} \{(z,x,p) : H^1(z,x,u,p) < a\},$$

we obtain that H is $\mathfrak{R}_{[0,T]^2} \times$ B $\times\, \mathfrak{R}_R$-measurable Therefore, there exists a $\mathfrak{R}_{[0,T]^2} \times B \times \mathfrak{R}_R$-measurable function $y^* : [0,T]^2 \times C[0,T]^2 \times R \to U$, which satisfies (4.9).

Next we define a measurable function $\psi(z,x)$,

$$\psi : \left([0,T]^2 \times C[0,T]^2, \mathfrak{R}_{[0,T]^2} \times \mathrm{B}\right) \to \left([0,T]^2 \times C[0,T]^2, \mathfrak{R}_{[0,T]^2} \times \mathrm{B} \times \mathfrak{R}_R\right).$$

Since $u^* = y^*(\psi)$ (implying that u^* is $\mathfrak{R}_{[0,T]^2} \times$ B-measurable), then $u^* \in$ U.

Let us prove that the control u^* is optimal. We have

$$Z(z) = J^* + \int_{[0,z]} (A(z') + B(z')a(z',x,u(z,x)))dz' + \int_{[0,z]} B(z')b(z')W(dz'),$$

where $(W(z), \mathfrak{I}_z, \mu_u)$ is a Wiener field. Put $z = (T,T)$ and compute the expectation with respect to μ_u:

$$J^* + E_u \int_{[0,z]} (A(z) + B(z)a(z,x,u(z,x)))dz = 0$$

From (4.8) we obtain that with P_u-probability 1

$$B(z)a(z,x,u(z,x)) + f(z,x,u(z,x)) \geq B(z)a(z,x,u^*(z,x)) + f(z,x,u^*(z,x)),$$

implying

$$J^* \leq E_u \int_{[0,T]^2} (f(z',x,u(z,x)))dz'$$
$$- E_u \int_{[0,T]^2} (A(z') - B(z')a(z',x,u^*(z,x)) - f(z',x,u^*(z,x)))dz'.$$

Thus, the control u^* is optimal whenever with probability 1

$$A(z') = -B(z')a(z',x,u^*(z,x)) + f(z',x,u^*(z,x)).$$

Let

$$X(x) = \int_{[0,T]^2} (A(z') - B(z')a(z',x,u^*(z,x)) - f(z',x,u^*(z,x)))dz',$$

where X is a nonnegative with probability 1 random variable. For every natural N define $X^N = X^N(\xi) = \min\{N, X(\xi)\}$. Since $Z((0,0),u) = J^*$, $Z(T,T) = 0$, then for every $\varepsilon > 0$ there exists $u \in \mathrm{U}$ such that

$$E_u \int_{[0,T]^2} (f(z,x,u(z,x)))dz < J^* + \varepsilon,$$

i.e. $E_u X < \varepsilon$. Therefore, there exists a sequence $\{u_n\} \subset \mathrm{U}$ such that $E_{u_n} X \to 0$, $n \to \infty$. Since $0 \leq X^N \leq X$, we have $E_{u_n} X^N \to 0$ as $n \to \infty$. Define

$$\varphi_n = \exp\left\{ \int_{[0,T]^2} a(z,x,u_n(z,x)) \left(\overline{b}^{-1}(z,x)\right)^2 x(dz) \right.$$
$$\left. -\tfrac{1}{2} \int_{[0,T]^2} a^2(z,x,u_n(z,x)) \left(\overline{b}^{-1}(z,x)\right)^2 dz \right\}.$$

For all n we have $E\varphi_n X^N \to 0$; on the other hand, since the set of densities φ_n is uniformly integrable and relatively weakly compact, $\varphi_n \to \varphi$ as $n \to \infty$. Therefore, we get $E\varphi X^N = 0$. Since $\varphi > 0$ and $\mu_\varphi \ll \mu$, we immediately get $X^N = 0$. Thus, the control u^* is optimal. □

Remark 4.2 Condition (4.8) holds, for example, if U is compact.

Remark 4.3 The uniqueness of the optimal control follows directly from Theorem 1.15.

Definition 4.5 The stochastic field $\xi(z)$ is a Markov field if for any set $B \in \mathcal{B}$

$$P\{\xi(T,T)\in B/\mathscr{B}_z^*\} = P\left\{\xi(T,T)\in B/\mathscr{B}_z^0\right\},$$

where $\mathscr{B}_z^0 = \sigma\{\xi_z(T,s), \xi_z(t,T), t \le T, s \le T\}$ and ξ_z is the solution to (4.1) in the domain $(z,(T,T)]$.

Remark 4.4 Since the solution to (4.1) (if it exists) defines a Markov field, see [30–32], therefore the optimal control is Markov as well.

4.2 Construction of an ε-Optimal Control for Diffusion-Type Random Fields

Now we discuss another principal problem appearing in stochastic control, namely, we derive the existence conditions of an ε-optimal control and describe the method of their construction.

We start with some auxiliary statements. Consider the sequence of random fields $(\xi_n(z))_{n\ge 1}$, $z \in [0,T]^2$, where $\xi_n(z)$ is defined as the solution to

$$\xi_n(z) = \int_{[0,z]} a(z_1, \xi_n(z_1), \eta_n(z_1))dz_1 + \int_{[0,z]} b(z_1, \xi_n(z_1))W(dz_1), \tag{4.10}$$

and $(\eta_n)_{n\ge 1}$ is the some sequence of random processes.

Lemma 4.5 *Assume that for almost all $z \in [0,T]^2$ we have the convergence of random processes $\eta_n(z) \to \eta_0(z)$. Then for any $\varepsilon > 0$*

$$E \sup_{z\in[0,T]^2} \left|\xi_n(t) - \xi(t)\right|^2 \to 0 \,\text{as}\quad n \to \infty.$$

The proof is analogous to the proof of Theorem 3.15 [26] and uses Lemma 4.1.

Definition 4.6 The control $\eta(z)$, $z \in [0,T]^2$, is called piecewise constant if there exists a set $\{z_{kj}\}$ such that $\eta(z) = \eta_{kj}$ for $z \in I_{kj} = (z_{kj}, z_{k+1,j+1}]$, where η_{kj} is an $\mathfrak{I}_{z_{kj}}$-measurable U-valued random variable.

Definition 4.7 Let $\mathbf{A}_1$ and $\mathbf{A}$ be two classes of admissible controls, $\mathbf{A}_1 \subset \mathbf{A}$. We say that the class $\mathbf{A}_1$ is dense in $\mathbf{A}$ if

$$\inf_{\eta\in\mathbf{A}_1} EF(\xi(\cdot),\eta(\cdot)) = \inf_{\eta\in\mathbf{A}} EF(\xi(\cdot),\eta(\cdot)).$$

In what follows we denote by $\mathbf{U}_\delta$ the class of piecewise constant controls corresponding to some given partition $\delta = \{z_{kj}\}$.

Theorem 4.3 *Suppose that the coefficients* a *and* b *satisfy the Lipschitz condition in x with some constant value K, and the cost function F(x,u) is bounded and continuous in the metric*

$$m(x_1,u_1,x_2,u_2) := \sup_{z\in[0,T]^2} |x_1(z)-x_2(z)| + \int_{[0,T]^2} |u_1(z)-u_2(z)|dz. \tag{4.11}$$

Then the class of all piecewise constant controls $\mathbf{U}_0$ *for equation (4.1) is dense in the class of all controls* $\mathbf{U}$.

Proof Let $\eta(z) = u(z,\omega)$, $z \in [0,T]^2$, be some control from $\mathbf{U}$. There exists a family of functions

$$u_n(z,\omega) = \sum_{k=1}^{n}\sum_{j=1}^{m} c_{kj}\chi_{I_{kj}(z)}\chi_{\Lambda_k}(\omega),$$

where I_{kj} was defined in Definition 4.5, $\Lambda_k \in \Im$, $c_{kj} \in U$, such that $u_n(z,\omega) \to u(z,\omega)$ as $n \to \infty$ $\lambda \times P$-almost everywhere, where λ is the Lebesgue measure on $[0,T]^2$. Since U is compact, then it is easy to see that

$$\int_{[0,T]^2}\int_{\Omega} |u_n(z,\omega) - u(z,\omega)|dzdP \to 0, n \to \infty.$$

Put $\widetilde{u}_n(z,\omega) = E[u_n(z,\omega)/\Im_z]$. Then the control $\widetilde{u}_n$ is piecewise constant, and

$$\begin{aligned}\int_{[0,T]^2}\int_{\Omega} |\widetilde{u}_n(z,\omega) - u(z,\omega)|dzdP &= \int_{[0,T]^2}\int_{\Omega} |E[\widetilde{u}_n(z,\omega) - u(z,\omega)/\Im_z]|dzdP \\ &\le \int_{[0,T]^2}\int_{\Omega} |u_n(z,\omega) - u(z,\omega)|dzdP \to 0, n \to \infty.\end{aligned}$$

By the Riesz theorem there exists a subsequence $\widetilde{u}_{n_k}$ converging to u $\lambda \times P$-almost everywhere. We construct the solutions ξ and ξ_k to (4.1), that correspond, respectively, to the controls u and $\widetilde{u}_{n_k}$. Since the cost function F is bounded, we have by Lemma 4.4

$$E\ F(\xi_k(\cdot), \widetilde{u}_{n_k}(\cdot)) \to E\ F(\xi(\cdot), u(\cdot)),$$

implying that $\inf_{u\in \mathbf{U}_0} F(\xi(\cdot),u(\cdot)) = \inf_{u\in\mathbf{U}} F(\xi(\cdot),u(\cdot))$. □

Remark 4.5 Let $\{u_1,u_2,\ldots,u_N\}$ be an arbitrary set of points from the space U. Denote by $U\{z_1,z_2,\ldots,z_n,u_1,u_2,\ldots,u_N\}$ some subset of the class $U\{z_1,z_2,\ldots,z_n\}$ of all piecewise constant controls under the partition $\{z_1,z_2,\ldots,z_n\}$. The set $U\{z_1, z_2,\ldots,z_n,u_1,u_2,\ldots,u_N\}$ consists of piecewise constant controls, with values in the set $\{u_1,u_2,\ldots,u_N\}$ only. Since the space U is compact, one can construct for it an ε-net for any $\varepsilon > 0$. Suppose that $\{u_1,u_2,\ldots,u_N\}$ is one of possible ε-nets. Clearly, for any function $u(z)$ from $U\{z_1,z_2,\ldots,z_n\}$ there exists a sequence $u_k(z)$, $k = 1, 2, \ldots$ from the set $U\{z_1,z_2,\ldots,z_n,u_1,u_2,\ldots,u_N\}$, that converges uniformly to $u(z)$. Thus, under the conditions of Theorem 4.2, for any countable everywhere dense sequence $\{u_1,u_2,\ldots,u_N,\ldots\}$ from U there exists an ε-optimal piecewise constant control from $U\{z_1,z_2,\ldots,z_n,u_1,u_2,\ldots,u_N\}$.

In the rectangle $[0,T]^2$ consider the partition $\delta = \left\{z_{kj}, k=\overline{0,n}, j=\overline{0,m}\right\}$, $|\delta| = \text{diam}(\delta)$ and choose some piecewise constant control η. For given δ and η we construct the random fields $\widehat{\xi}_\delta$ and $\breve{\xi}_\delta$ as below:

$$\begin{aligned}
&\breve{\xi}_\delta = \xi_{kj}, z\in I_{kj} = \left(z_{kj}, z_{k+1,j+1}\right],\\
&\xi_{k+1,j} = \xi_{kj} + \sum_{l=0}^{k-1}\left(\int_{\left[z_{lj},z_{l+1,j+1}\right]} a\left(z, \breve{\xi}_\delta, \eta_{kj}\right)dz + \int_{\left[z_{lj},z_{l+1,j+1}\right]} b\left(z, \breve{\xi}_\delta\right)W(dz)\right),\\
&\xi_{k,j+1} = \xi_{kj} + \sum_{l=0}^{j-1}\left(\int_{\left[z_{kl},z_{k+1,l+1}\right]} a\left(z, \breve{\xi}_\delta, \eta_{kj}\right)dz + \int_{\left[z_{kj},z_{k+1,l+1}\right]} b\left(z, \breve{\xi}_\delta\right)W(dz)\right).
\end{aligned}$$

Let

$$\widehat{\xi}_\delta(z) = \int_{[0,z]} a\left(z', \breve{\xi}_\delta, \eta(z')\right)dz' + \int_{[0,z]} b\left(z', \breve{\xi}_\delta\right)W(dz').$$

Obviously, these random fields depend on δ and η. Denote by ξ_δ the solution to (4.1), corresponding to the control η_δ, and consider the cost function $F(u)$, defined in (4.2).

Lemma 4.6 *Suppose that the conditions of Theorem 4.2 are satisfied. Then* $\lim\limits_{|\delta|\to 0} \left|F\left(\breve{\xi}_\delta, \eta_\delta\right) - F(\xi_\delta, \eta_\delta)\right| = 0$, *uniformly in all piecewise constant controls* $\eta_\delta \in U_\delta$.

Proof Since $F\left(\breve{\xi}_\delta, \eta_\delta\right) - F(\xi_\delta, \eta_\delta) \to 0$ in probability as $|\delta| \to 0$, and the function $F_0(x,\eta_\delta)$ is continuous in x, the necessity follows from the construction for $\breve{\xi}_\delta$.

For a given δ consider the sequence (ξ_{kj},η_{kj}) and define the cost function

$$F_\delta(x,u) = F\left(\breve{x}_\delta, \breve{u}_\delta\right)\breve{x}_\delta(z) = x_{kj}\breve{x}_\delta(z) = x_{kj}, \breve{u}_\delta(z) = u_{kj}, z\in I_{kj}. \tag{4.12}$$

It is easy to see that there exists a sequence $(\eta_{kj}{}^*)$ which minimizes the cost function EF_δ. Put $\eta_\delta{}^*(z) = \eta_{kj}{}^*$, $z \in I_{kj}$. One can show that for sufficiently small $|\delta|$ the control $\eta_\delta{}^*(z)$ is ε-optimal for the cost function F, which finished the proof of the lemma.

Theorem 4.4 *Let* a, $b \in S(C,L)$, *with the respective constants independent of* u *and* z. *Suppose that the cost function* F(x,u) *is bounded and continuous in the metric defined in (4.11). Then*

$$Z = \inf_{\eta \in \mathbf{U}} EF_0(\xi_\eta, \eta) = \lim_{|\delta| \to 0} EF_\delta\left(\xi^*_{kj}, \eta^*_{kj}, k = \overline{0,n}, j = \overline{0,m}\right).$$

Proof Since $F_\delta\left(\xi^*_{kj}, \eta^*_{kj}\right) = F\left(\breve{\xi}^*_\delta, \eta^*_\delta\right)$, where $\breve{\xi}^*_\delta$ is a piecewise-constant approximation of the solution to (4.1), constructed from the control $\eta_\delta{}^*$, we have $Z \leq \lim_{|\delta| \to 0} EF\left(\breve{\xi}^*_\delta, \eta^*_\delta\right)$. At the same time, by Theorem 4.2 for any $\varepsilon > 0$ there exists a piecewise-constant control $\eta_\varepsilon(z)$ such that $EF(\xi_\varepsilon,\eta_\varepsilon) < Z + \varepsilon$, where $\xi_\varepsilon(z)$ is the solution to (4.1), corresponding to the control η_ε. Consider all possible partitions of $[0,T]^2$ which are the sub-partitions of the partition δ. Let $\eta_\varepsilon(z) = \eta_{\delta'}(z)$, then for every $z \in [0,T]^2$ we have $\xi_\varepsilon(z) = \xi_{\delta'}(z)$ and $\lim_{|\delta| \to 0} EF\left(\breve{\xi}_{\delta'}, \eta_{\delta'}\right) = F(\xi_\varepsilon, \eta_\varepsilon)$. Thus, for $|\delta'|$ sufficiently small we obtain $F\left(\breve{\xi}_{\delta'}, \eta_{\delta'}\right) < Z + \varepsilon$ and, consequently,

$$EF_{\delta'}\left(\xi^*_{kj}, \eta^*_{kj}, k = \overline{0,n}, j = \overline{0,m}\right) < Z + \varepsilon,$$

which in turn implies

$$\overline{\lim_{|\delta'| \to 0}}\ EF_{\delta'}\left(\xi^*_{kj}, \eta^*_{kj}\right) \leq Z. \qquad \square$$

To make the presentation self-contained we also prove the Bellman principle for processes defined on the plane. Let $I_z = (t, T] \times (s, T]$. We can write the optimal cost function on the set I_z, introduced in (4.5), as

$$F_{I_z}(x, u) = \int_{I_z} f(z', x, u) dz',$$

provided that $\xi_z{}^u(z') = x(z')$, $z' \in [0,T]^2 \backslash I_z$, where $\mathbf{U}_0(I_z)$ is the class of piecewise constant controls on the rectangle I_z.

Theorem 4.5 *Let the cost function be given by (4.2). Then for all* $z_1 \in [0,T]^2$, $z_1 \geq z$, *we have*

$$Z(z, x) = \inf_{u \in \mathbf{U}_0\left(I_z \backslash I_{z_1}\right)} E\left[F_{I_z \backslash I_{z_1}}\left(\xi^u_z, u\right) + Z\left(z_1, \xi^u_z\right)\right]. \tag{4.13}$$

Proof Fix z, x, and some $\varepsilon > 0$. For such ε there exist a partition δ and a control $\eta \in \mathbf{U}_\delta$, such that

$$\begin{aligned} Z(z, x) + \varepsilon > EF_{I_z}\left(\xi^\eta_z, \eta\right) &= E\left[F_{I_z \backslash I_{z_1}}\left(\xi^\eta_z, \eta\right) + F_{I_{z_1}}\left(\xi^\eta_z, \eta\right)\right] \\ &\geq E\left[F_{I_z \backslash I_{z_1}}\left(\xi^\eta_z, \eta\right) + Z\left(z_1, \xi^\eta_z\right)\right], \end{aligned}$$

implying that $Z(z, x) \geq E\left[F_{I_z \backslash I_{z_1}}\left(\xi^\eta_z, \eta\right) + Z\left(z_1, \xi^\eta_z\right)\right]$.

Let us prove the sufficiency of the reverse inequality. Let η^* be an optimal control for the sequence ξ_{kj}^{η} which corresponds to the partition δ. If for $\|x\|_z \le N$ we have $|\delta| \to 0$, then the expectation $EF_{I_z}(\xi_z^{\eta^*}, \eta^*)$ converges to $Z(z,x)$ uniformly in x. Moreover, the random variable $\sup_{z_1 \in [0,T]^2} |\xi_z^{\eta}(z_1)|$ is stochastically bounded, implying that there exists some δ, such that

$$P\left\{E\left[F_{I_{z_1}}(\xi_z^{\eta^*}, \eta^*)/\mathfrak{I}_{z_1}^*\right] < Z\left(z_1, \xi_z^{\eta^*}(z_1)\right) + \varepsilon\right\} > 1 - \varepsilon.$$

We obtain

$$\begin{aligned} Z(z,x) &= \inf_{u \in \mathrm{U}_0(I_z I_{z_1})} EF_{I_z}(\xi_z^u, u) \\ &= \inf_{u \in \mathrm{U}_0(I_z I_{z_1})} \left\{EF_{I_z \setminus I_{z_1}}(\xi_z^u, u) + E\left[F_{I_{z_1}}(\xi_z^{\eta^*}, \eta^*)/\mathfrak{I}_{z_1}^*\right]\right\}. \end{aligned}$$

Since each control $\eta \in \mathbf{U}(I_z)$ can be constructed from two components $\eta_1 \in \mathbf{U}(I_z \setminus I_{z_1})$ and $\eta_2 \in \mathbf{U}(I_{z_1})$, we obtain

$$\begin{aligned} Z(z,x) &\le \inf_{u \in \mathrm{U}_0(I_z I_{z_1})} EF_{I_z}(\xi_z^u, u) + \left(Z\left(z_1, \xi_z^{\eta^*}(z_1)\right) + \varepsilon + c\varepsilon\right) \\ &\le c_1\varepsilon + \inf_{u \in \mathrm{U}_0(I_z I_{z_1})} EF_{I_z}(\xi_z^{\eta}, \eta) + Z\left(z_1, \xi_z^{\eta^*}(z_1)\right). \end{aligned}$$

Taking $\varepsilon > 0$ small enough, we get

$$Z(z,x) \le \inf_{u \in U_0(I_z I_{z_1})} EF_{I_z}(\xi_z^u, u) + Z(z_1, \xi u(z_1)). \qquad \square$$

Let us now proceed with the construction of an ε-optimal control. Since the function $f(z,x,u)$ is continuous, we can assume that the cost function F_δ is given by

$$F_\delta(x,u) = \sum_{k,j} f\left(z_{kj}, x(z_{kj}), u(z_{kj})\right)(t_{k+1} - t_k)\left(s_{j+1} - s_j\right). \tag{4.14}$$

The optimal control for the solution to (4.1) with the cost function (4.14) can be defined as follows. First we construct the sequence of functions $Z_{kj}(x)$, defined below:

$$Z_{kj}(x) = \min_u Z_{kj}^*(x,u),$$

$$Z_{n+1,l}^*(x,u) = Z_{i,m+1}^*(x,u) = 0, l = \overline{0, m+1}, i = \overline{0, n+1},$$

$$Z_{kj}^{*}(x,u) = f(z_{kj},x,u)(t_{k+1}-t_k)(s_{j+1}-s_j) + E \sum_{\substack{k \le i \le n+1 \\ j \le l \le m+1 \\ (i,l) \neq (k,j)}} Z_{il}\Big(\xi_{(t_i,s_l)}(z_{kj},x,u)\Big),$$

where

$$\xi_{(t,s)}(z_{kj},x,u) = x + \int_0^t \int_{s_j}^s a(z',x,u)dz' + \int_{t_k}^t \int_0^{s_j} a(z',x,u)dz' + \int_0^t \int_{s_j}^s b(z',x)W(dz') + \int_{t_k}^t \int_0^{s_j} b(z',x)W(dz').$$

Denote by $g_{kj}(x)$ the Borel function for which $Z_{kj}(x) = Z_{kj}^{*}(x, g_{kj}(x))$. Clearly, the control $u = g_\delta(z,x) = g_{kj}(x)$, $z \in I_{kj}$ is ε-optimal for sufficiently small $|\delta|$.

Now let $\eta(z) = \varphi(z, \xi(z))$ be a Markov control for the solution to (4.1), such that for all $z \in [0,T]^2$, $l = k-1, k$, $i = j-1, j$, $(i,l) \neq (k,j)$,

$$Ef\Big(z_{kj}, \xi^{\varphi}_{z_{li},x}(z_{kj}), \eta(z_{kj})\Big) \le \inf_u f\Big(z_{kj}, \xi_{z_{kj}}(z_{li},x,u),u\Big). \tag{4.15}$$

Then

$$Ef\Big(z_{kj}, \xi^{\varphi}_{0,x}(z_{kj}), \eta(z_{kj})\Big) \le Ef\Big(z_{kj}, \xi^{\varphi}_{z_{lj}}\Big(z_{li}, \xi^{\varphi}_{0,x}(z_{li}),u\Big),u\Big),$$

and also

$$Ef\Big(z_{kj}, \xi^{\varphi}_{0,x}(z_{kj}), \eta(z_{kj})\Big) \le Ef\Big(z_{kj}, \xi_{kj}, g_{kj}(\xi_{kj})\Big),$$

where the sequence (ξ_{kj}) is defined as follows:

$$\xi_0 = x, \xi_{k+1,j} = \xi_{k+1,j}\Big(z_{kj}, \xi_{z_{kj}}, g_{kj}\Big(\xi_{z_{kj}}\Big)\Big), \xi_{k,j+1} = \xi_{k+1,j+1}\Big(z_{kj}, \xi_{z_{kj}}, g_{kj}\Big(\xi_{z_{kj}}\Big)\Big).$$

Therefore, to construct an ε-optimal Markov control it is sufficient to construct such a control, for which (4.15) is satisfied. To do this, it is necessary to specify for each I_{kj} the corresponding function $\varphi(z,x)$. We assume that the space U consists of finite number of points $u_1, \ldots, u_N$, and instead of I_{kj} consider the rectangle $[0,T]^2$. Put $f(z_{kj},x,u) =: h(x,u)$.

Let us construct the necessary control following the scheme presented below. Let $Z(z,x) = \min_u Eh\Big(\xi_{(T,T)}(z,x,u),u\Big)$, and assume that $\varphi(z,x)$ satisfies the equation

$$Z(z,x) = Eh\Big(\xi_{(T,T)}(z,x,\phi(z,x)),\phi(z,x)\Big).$$

For some partition δ of $[0,T]^2$ by the set of points z_{kj} we denote by $\eta(z)$ the piecewise constant control with $\eta(z_{kj}) = \varphi(z_{kj}, \xi(z_{kj}))$. Let $\breve{\xi}_{z_{kj},x}(z)$ be the solution to (4.1) on the rectangle I_{kj}, corresponding to the control $\eta(z)$. Then

$$Z(z_{kj},x) = Eh\left[\xi_{(T,T)}(z_{kj},x,\varphi(z_{kj},x)),\varphi(z_{kj},x)\right]$$
$$= Eh\left[\breve{\xi}_{z_{kj},x}(T,T),\varphi\left(z_{kj},\breve{\xi}_{z_{kj},x}(z_{kj})\right)\right]$$

$$\begin{aligned}
Z(z_{k-1,j},x) &= Eh\left[\xi_{(T,T)}(z_{k-1,j},x,\varphi(z_{k-1,j},x)),\varphi(z_{k-1,j},x)\right]\\
&= Eh\left[\xi_{(T,T)}\left(z_{k-1,j},\breve{\xi}_{z_{k-1,j},x}(z_{kj},\varphi(z_{k-1,j},x),\varphi(z_{k-1,j},x)\right),\varphi(z_{k-1,j},x)\right]\\
&\geq E\inf_u E\left[h\left(\xi_{(T,T)}\left(z_{k-1,j},\breve{\xi}_{z_{k-1,j},x}(z_{kj},u),u\right)\right)/\Phi^*_{kj}\right]\\
&= Eh\left[\xi_{(T,T)}\left(z_{k-1,j},\breve{\xi}_{z_{k-1,j},x}(z_{kj},\varphi(z_{k-1,j},x)),\varphi(z_{k-1,j},x)\right)\right]\\
&\geq Eh\left[\xi_{(T,T)}\left(z_{k-1,j},\breve{\xi}_{z_{k-1,j},x}(z_{kj}),\varphi\left(z_{k-1,j},\breve{\xi}_{z_{k-1,j},x}(z_{kj})\right)\right),\right.\\
&\left.\varphi\left(z_{k-1,j},\breve{\xi}_{z_{k-1,j},x}(z_{kj})\right)\right] = Eh\left(\breve{\xi}_{z_{k-1,j},x}(T,T),\phi\left(z_{kj},\breve{\xi}_{z_{k-1,j},x}(z_{kj})\right)\right),
\end{aligned}$$

where Φ_{kj}^* is the σ-algebra with respect to which $\xi(t,s_{k-1}) - \xi(t_{k-1},s_{k-1})$ and $\xi(t_{k-1},s) - \xi(t_{k-1},s_{k-1})$ are measurable for any $z = (t,s) \in I_{kj}$ and on the borders: for $z = (t,s_j)$, $t \in [t_k,T]$, $z = (t_k,s)$, $s \in [s_j,T]$. Similar equality takes place also for $Z(z_{k,j-1},x)$. The above relations give

$$Eh\left(\breve{\xi}_{0,x}(T,T),\eta(T,T)\right) \leq \inf_u h\left(\breve{\xi}_{(T,T)}(0,x,u),u\right).$$

If the finite-dimensional distributions of the fields $\breve{\xi}_{0,x}(z)$ and $\eta(z)$ converge as $|\delta| \to 0$ to finite-dimensional distributions of some stochastic fields $\xi_{0,x}(z)$ and $\eta(z)$, then

$$Eh(\xi_{0,x}(T,T),\eta(T,T)) \leq \inf_u h(\xi(0,x,u),u).$$

Further, if $\eta(z) = \varphi(z, \xi(z))$ with probability 1 for almost all $z \in [0,T]^2$, and $\xi_{0,x}(z)$ is the solution to

$$\xi_{0,x}(z) = \xi_0 + \int_{[0,z]} a(z',\xi_{0,x}(z'),\eta(z'))dz' + \int_{[0,z]} b(z',\xi_{0,x}(z'))W(dz'),$$

then the constructed control is ε-optimal.

Consider the field $\xi^\delta(z)$ satisfying the equation

$$\xi^{\delta}(z) = \xi^{\delta}(z_{kj}) + \sum_{l=1}^{N} \chi_{J_l}(z_{kj}, \xi^{\delta}(z_{kj}))$$
$$\times \left\{ \int_0^t \int_{s_j}^s a_1(z', \xi^{\delta}(z'))dz' + \int_0^t \int_{s_j}^s b(z', \xi^{\delta}(z'))W(dz') + \right.$$
$$\left. + \int_{t_k}^t \int_0^s a_1(z', \xi^{\delta}(z'))dz' \int_{t_k}^t \int_0^s b(z', \xi^{\delta}(z'))W(dz') \right\},$$
$$z = (t,s) \in S_{kj} = \left(\bigcup_{p=0}^{j} I_{kp} \right) \cup \left(\bigcup_{i=0}^{k} I_{ij} \right),$$
$$J_l = \{(z,x) : \varphi(z,x) = u_1\}, a_1(z,x) = a(z,x,u_1).$$

Denote by μ_δ the measure on the space $(C[0,T]^2, \mathscr{B})$ corresponding to the field $\xi^{\delta}(z)$ and recall (from Sect. 4.1) that μ is the measure on the space $(C[0,T]^2, \mathscr{B})$ corresponding to the solution of the equation

$$x(z) = x_0 + \int_{[0,z]} b(z_1, x(z_1))W(dz_1)$$

Measures μ_δ are absolutely continuous with respect to μ and the density $\frac{d\mu_\delta}{d\mu} = \zeta_\delta(T,T)$ admits the representation

$$\zeta_\delta(T,T) = \exp\left\{ \int_{[0,T]^2} \left(\overline{b}^{-1}(z,x)\right)^2 A_\delta(z,x)x(dz) - \frac{1}{2} \int_{[0,T]^2} \left(\overline{b}^{-1}(z,x)A_\delta(z,x)\right)^2 dz \right\},$$

where $A_\delta(z,x) = \sum_{l=1}^{N} \chi_{J_l}(z_{kj}, x(z_{kj}))a_1(z, x(z)), z \in S_{kj}$. Let us show that the family of densities ζ_δ is uniformly integrable with respect to μ. Since $\left|\overline{b}^{-1} A_\delta\right| \leq c$, we have

$$\int_{C[0,T]^2} \zeta_\delta^2(T,T)d\mu$$
$$= \int_{C[0,T]^2} \exp\left\{ 2\int_{[0,T]^2} \left(\overline{b}^{-1}(z,x)\right)^2 A_\delta(z,x)x(dz) - \int_{[0,T]^2} \left(\overline{b}^{-1}(z,x)A_\delta(z,x)\right)^2 dz \right\} d\mu$$
$$\leq c_1 \int_{C[0,T]^2} \exp\left\{ 2\int_{[0,T]^2} \left(\overline{b}^{-1}(z,x)\right)^2 A_\delta(z,x)x(dz) - 2\int_{[0,T]^2} \left(\overline{b}^{-1}(z,x)A_\delta(z,x)\right)^2 dz \right\} d\mu \leq c_1.$$

Lemma 4.7 *Let*
$|b(z,x) - b(z',x')| \leq c(|z - z'| + |x - x'|)$ *for some constant* c

and

$$\sup_{z,x} \left(|a(z,x,u)| + |b(z,x)| + \left|\overline{b}^{-1}(z,x)\right| \right) < \infty.$$

Then

$$\lim_{|\delta|\to 0} E\int_{[0,T]^2} |a(z,x(z)) - A_\delta(z,x(z))|^2 dz = 0.$$

Proof It is easy to see that

$$\begin{aligned} &E\int_{[0,T]^2} |a(z,x(z)) - A_\delta(z,x(z))|^2 dz \\ &= \sum_{k=0}^{n}\sum_{j=0}^{m} E\int_{I_{kj}} \left|\sum_{l=1}^{N} \chi_{J_l}(z_{kj},x(z_{kj})) - \chi_{J_l}(z,x(z))a_1(z,x(z))\right|^2 dz \\ &\le c_N \sum_{l=1}^{N}\sum_{k=0}^{n}\sum_{j=0}^{m} E\int_{I_{kj}} \left(\chi_{J_l}(z_{kj},x(z_{kj})) - \chi_{J_l}(z,x(z))a_1(z,x(z))\right)^2 dz. \end{aligned}$$

Therefore, it suffices to prove that for any l

$$B(l,\delta) = \sum_{k=0}^{n}\sum_{j=0}^{m} E\int_{I_{kj}} \left(\chi_{J_l}(z_{kj},x(z_{kj})) - \chi_{J_l}(z,x(z))a_1(z,x(z))\right)^2 dz \to 0$$

as $|\delta| \to 0$. Since J_l is the Borel set, then for any $\varepsilon > 0$ there exist two closed sets J^1 and J^2, such that $J^1 \subset J_l, J^2 \subset [0,T]^2 \times C[0,T]^2 \backslash J_l$, with $\lambda \times \mu\left(\overline{J^2} \backslash J^1\right) < \varepsilon$, $\overline{J^2} = [0,1]^2 \times C[0,T]^2 \backslash J^2$.

Let $\psi(z,x)$ be a continuous function, $0 \le \psi(z,x) \le 1$,

$$\psi(z,x) = \begin{cases} 0, & (z,x)\in J^2, \\ 1, & (z,x)\in J^1. \end{cases}$$

Then

$$\begin{aligned} B(l,\delta) &= \sum_{k=0}^{n}\sum_{j=0}^{m}\left(B_{kj}^{(1)} + B_{kj}^{(2)} + B_{kj}^{(3)}\right) \\ B_{kj}^{(1)} &= \int_{I_{kj}}\int_{C[0,T]^2} \left(\chi_{J_l}(z_{kj},x(z_{kj})) - \psi(z_{kj},x(z_{kj}))\right)^2 d\mu(x)dz, \\ B_{kj}^{(2)} &= \int_{I_{kj}}\int_{C[0,T]^2} \left(\psi(z,x(z)) - \psi(z_{kj},x(z_{kj}))\right)^2 d\mu(x)dz, \\ B_{kj}^{(3)} &= \int_{I_{kj}}\int_{C[0,T]^2} \left(\chi_{J_l}(z,x(z)) - \psi(z,x(z))\right)^2 d\mu(x)dz. \end{aligned}$$

By continuity of ψ we get

$$\sum_{k=0}^{n}\sum_{j=0}^{m} B_{kj}^{(2)} \to 0, |\delta| \to 0$$

Let us construct a continuous function $g(z,x)$ such that $0 \le g(z,x) \le 1$, $g(z,x) = 1$ for $(z,x) \in \overline{J^2} \setminus J^1$ and $\lambda \times \mu\{(z,x) : g(z,x) > 0\} < 2\varepsilon$. Then

$$\left(\chi_{J_l}(z_{kj}, x(z_{kj})) - \psi(z_{kj}, x(z_{kj}))\right)^2 \le g(z,x),$$

and

$$\overline{\lim_{|\delta|\to 0}} \sum_{k=0}^{n} \sum_{j=0}^{m} B_{kj}^{(1)} \le \sum_{k=0}^{n} \sum_{j=0}^{m} \lambda(I_{kj}) \int_{C[0,T]^2} g(z_{kj}, x) d\mu(x)$$
$$\le c_0 \lambda \times \mu\{(z,x) : g(z,x) > 0\} < 2\varepsilon.$$

Similarly, one can derive that $\overline{\lim}_{|\delta|\to 0} \sum_{k=0}^{n} \sum_{j=0}^{m} B_{kj}^{(3)} < 2\varepsilon$. □

Theorem 4.6 *Suppose that functions a(z,x,u), b(z,x) satisfy local Lipschitz condition,* $|b(z,x) - b(z',x')| \le c(|z - z'| + |x - x'|)$ *for some constant c and*

$$\sup_{z,x} \left(|a(z,x,u)| + |b(z,x)| + \left|\overline{b}^{-1}(z,x)\right| \right) < \infty.$$

Then the finite-dimensional distributions of $\xi^\delta(z)$ *converge weakly as* $|\delta| \to 0$ *to the finite-dimensional distributions of the stochastic field satisfying the equation*

$$\xi(z) = \xi_0 + \int_{[0,z]} a(z', \xi(z'), u) dz' + \int_{[0,z]} b(z', \xi(z')) W(dz'). \tag{4.16}$$

Proof We have by Lemma 4.7 that

$$E\left\{ \int_{[0,T]^2} (b^{-1}(z,x))^2 A_\delta(z,x) x(dz) - \int_{[0,T]^2} \left(\overline{b}^{-1}(z,x)\right)^2 a(z,x) x(dz) \right\}^2$$
$$\le c^2 E \int_{[0,T]^2} |a(z,x(z)) - A_\delta(z,x(z))|^2 dz \to 0, |\delta| \to 0.$$

Moreover,

$$E\left| \int_{[0,T]^2} \left(\overline{b}^{-1}(z,x) a(z,x)\right)^2 dz - \int_{[0,T]^2} \left(\overline{b}^{-1}(z,x) A_\delta(z,x)\right)^2 dz \right|$$
$$\le 2c^2 \left(E \int_{[0,T]^2} |a(z,x(z)) - A_\delta(z,x(z))|^2 dz \right)^{1/2} \to 0, |\delta| \to 0.$$

From the above relations we derive that $\lim_{|\delta|\to 0} \zeta_\delta(T,T) = \zeta(T,T)$, where

$$\zeta(T,T) = \exp\left\{\int_{[0,T]^2} \left(\overline{b}^{-1}(z,x)\right)^2 a(z,x)x(dz) - \frac{1}{2}\int_{[0,T]^2} \left(\overline{b}^{-1}(z,x)a(z,x)\right)^2 dz\right\}.$$

Since the set of densities $\zeta_\delta(T,T)$ is uniformly integrable, we get

$$\lim_{|\delta|\to 0} \int_{C[0,T]^2} f(x)d\mu_\delta = \lim_{|\delta|\to 0} \int_{C[0,T]^2} f(x)\zeta_\delta(T,T)d\mu = \int_{C[0,T]^2} f(x)\zeta(T,T)d\mu,$$

which in turn implies that the sequence of measures μ_δ converges weakly as $|\delta| \to 0$ to the measure $\widetilde{\mu}$, which is absolutely continuous with respect to μ with the density $\zeta(T,T)$. But $\widetilde{\mu}$ corresponds to the solution to (4.1). □

Corollary 4.3 *Under assumptions E1–E6 for any ε > 0 there exists an ε-optimal piecewise control for the field given by (4.1) and cost function (4.2).*

Corollary 4.4 *Suppose* $\zeta(z) = E\{\zeta(T,T)/\mathcal{B}_z\}$, $z \in [0,1]^2$. *Then* $\left\{\zeta(z), C[0,T]^2, \mathcal{B}_z, \mu\right\}$ *is the continuous nonnegative martingale.*

Chapter 5
Stochastic Processes in a Hilbert Space

In this chapter we consider essential problems of stochastic processes with values in a Hilbert space. We present an analogue of the Girsanov theorem for processes of such a type, and some filtration and optimal control problems. Results, exposed in Sects. 5.1 and 5.2, are published in [70], results of Sect. 5.3 are published in [49, 50].

Let H be a real separable Hilbert space with the scalar product (h_1,h_2), $h_1, h_2 \in H$, and a norm $|h|$, $h \in H$. In what follows, any process is assumed to take values either in H, or in some extension of H. Our main object is a Wiener process on H. This process can be naturally interpreted as a stochastic process whose coordinates in some orthonormal basis are one-dimensional Wiener processes. It is clear that this process cannot be considered in the space of functions taking values in H. However, see [9, 28], it can be regarded as a process on the space H_-, which is some extension of H with respect to the norm $|h|_- = |K^{1/2}h|$, where K is an arbitrary self-adjoint positive definite operator on H.

Denote by H_+ the Hilbert space obtained from the domain of operator $K^{-1/2}$ by taking the scalar product $(h_1,h_2)_+ = (K^{-1/2}h_1, h_2K^{-1/2})$. One usually refers to the triple $H_+ \subset H \subset H_-$ as to the "equipped Hilbert space", or the "rigged Hilbert space."

5.1 Ito Processes and Diffusion-Type Processes in a Hilbert Space

Let $(\Omega, \Im, P)$ be a probability space. We denote by $W = (W(t), \Im_t, P)$ the standard Wiener process on $(\Omega, \Im, P)$ taking values in H_-, where $(\Im_t, t\in[0,T])$ is a filtration such that for any $t \in [0,T]$ the random variable $W(t)$ is $\Im_t$-measurable, and $w(u) - w(t)$ does not depend on $\Im_t$, $0 \le t < u \le T$.

Further we assume that we have a fixed orthonormal basis (e_i) in H, consisting of eigenvectors of the operator W. In this basis $W(t)$ can be decomposed as

P.S. Knopov and O.N. Deriyeva, *Estimation and Control Problems for Stochastic Partial Differential Equations*, Springer Optimization and Its Applications 83,
DOI 10.1007/978-1-4614-8286-4_5,

$$W(t) = \sum_{i=1}^{\infty} W_i(t) e_i,$$

where $(W_i(t), \Im_t, P)$ are independent one-dimensional Wiener processes.

The notion of a stochastic integral with respect to a Wiener process can be introduced similarly to those in the finite-dimensional case. Denote by $L_2(H, \Im_t)$ (respectively, by $L(H, \Im_t)$) the class of $\Im_t$-adapted stochastic processes $\varphi = (\varphi(t), t \in [0,T])$ which take values in H, and satisfy

$$\int_0^T E|\varphi(t)|^2 dt < \infty$$

(respectively, $P\left\{ \int_0^T |\varphi(t)|^2 dt < \infty \right\} = 1$).

Let $L_2(L_2(H,H), \Im_t)$ be the class of all $\Im_t$-adapted random processes $B = (B(t), t \in [0,T])$ with values in $L_2(H,H)$, which is the space of Hilbert–Schmidt operators on H such that

$$\int_0^T E\|B(t)\|_-^2 dt < \infty,$$

where $\|\cdot\|_-$ is the operator norm in $L_2(H,H)$.

For functions belonging to the classes $L_2(H, \Im_t)$ and $L(H, \Im_t)$ the stochastic integral is defined as follows [28, 67]:

$$I_t(\varphi) = \int_0^T (\varphi(t), dw(t)), \quad \varphi \in L_2(H, \Im_t) \text{ or } \varphi \in L(H, \Im_t)$$

$$I_t(B) = \int_0^T B(t) dw(t), \quad B \in L_2(L_2(H,H), \Im_t) \text{ or } B \in L_2(L(H_-, H_-), \Im_t).$$

We refer to [70] for the proofs of the properties of stochastic integral, presented below:

1. If $\varphi \in L_2(H, \Im_t)$ and $B \in L_2(L_2(H,H), \Im_t)$ or $B \in L_2(L(H_-, H_-), \Im_t)$, then the processes $(I_t(\varphi), t \in [0,T])$ and $(I_t(B), t \in [0,T])$ are square integrable (with appropriate norms) martingales with values in R, H, and H_-, respectively, and their separable modifications are continuous.

2. $EI_t(\varphi) = 0,\ t \in [0,T],\ \varphi \in L_2(H, \Im_t)$;

$$EI_t(B) = 0, \quad t \in [0,T], \ B \in L_2(L_2(H,H), \Im_t) \text{ or } B \in L_2(L(H_-,H_-), \Im_t).$$

3. $EI_t^2(\varphi) = \int_0^t E|\varphi(\tau)|^2 d\tau,\ t \in [0,T],\ \varphi \in L_2(H, \Im_t)$;

4. $E|I_t(B)|^2 = \int_0^t E\|B(\tau)\|_2^2 d\tau,\ t \in [0,T],\ B \in L_2(L_2(H,H), \Im_t)$;

5. $E|I_t(B)|_-^2 = \int_0^t E\|B(\tau)\|_-^2 d\tau,\ t \in [0,T],\ B \in L_2(L(H_-,H_-), \Im_t)$.

Denote by C_T the Banach space of continuous functions $x = \{x(t),\ t \in [0,T]\}$ taking values in H_-, and endowed it with the norm $\|x\|_- = \sup_{0 \le t \le T} |x(t)|_-$. Further, denote by $\Re_t$ the σ-algebra of subsets generated by cylinders with bases in $[0,t]$, $t \in [0,T]$

Definition 5.1 ([33, 54]) An $\Im_t$-adapted process $\xi = (\xi(t), \Im_t)$ on $(\Omega, \Im, P)$ with values in H_- is called a diffusion-type process, if there exists a measurable functional $\alpha(t,x)$, where $t \in [0,T]$, $x \in C_T$, independent of the future (i.e., is $\Re_t$-measurable for any t) and with values in H, such that

$$P\left\{ \int_0^T |\alpha(t,\xi)| dt < \infty \right\} = 1,$$

and for any $t \in [0,T]$

$$\xi(t) = \xi(0) + \int_0^t \alpha(\tau, \xi) d\tau + W(t), \quad P\text{-a.s.} \tag{5.1}$$

Here $\alpha(t,\xi) = \alpha(t, \xi(\omega))$.

Definition 5.2 ([33, 54]) A continuous $\Im_t$-adapted stochastic process $\xi = (\xi(t), \Im_t)$ on $(\Omega, \Im, P)$ with values in H_- is called an Ito process, if there exists a random process $(\alpha(t), \Im_t)$ with values in H, such that

$$P\left\{ \int_0^T |\alpha(t)| dt < \infty \right\} = 1,$$

and for any $t \in [0,T]$

$$\xi(t) = \xi(0) + \int_0^t \alpha(\tau)d\tau + w(t), \quad P\text{-a.s.} \tag{5.2}$$

Analogously, one can define an Ito diffusion, as well as diffusion-type processes with the diffusion operator not necessarily equal to I:

$$\xi(t) = \xi(0) + \int_0^t \alpha(\tau,\xi)d\tau + \int_0^t B(\tau,\xi)dW(\tau),$$

$$\xi(t) = \xi(0) + \int_0^t \alpha(\tau)d\tau + \int_0^t B(\tau)dW(\tau),$$

where $B \in L_2(L_2(H,H),\Im_t)$ or $B \in L_2(L(H_-,H_-),\Im_t)$.

Further, we investigate diffusion-type and Ito processes of the form (5.1) and (5.2), with $\xi(0) = 0$ P-a.s.

Let μ_ξ and μ_w be the measures in the space $(C_T, \Re_T)$, generated by ξ and W, respectively. Put

$$\zeta_\tau^t(\varphi) = \exp\left\{ \int_\tau^t (\varphi(u), dW(u)) - \frac{1}{2}\int_\tau^t |\varphi(u)|^2 du \right\}, \quad 0 \le \tau < t \le T,$$

$$\varphi = (\varphi(t), \Im_t) \in L_2(H, \Im_t).$$

As in the one-dimensional case, one can formulate the respective statements in the Hilbert-space setting.

Lemma 5.1 *For any $\varphi = (\varphi(t),\Im_t) \in L_2(H,\Im_t)$ the process $\left(\zeta_0^t(\varphi),\Im_t\right)$ is a supermartingale, i.e. $E\left[\zeta_0^t(\varphi)/\Im_\tau\right] \le \zeta_0^\tau(\varphi)$ P-a.s., $0 \le \tau < t \le T$.*

Lemma 5.2 *If $\varphi = (\varphi(t),\Im_t) \in L_2(H,\Im_t)$ and*

$$E\zeta_0^T(\varphi) = 1, \tag{5.3}$$

then the process $\left(\zeta_0^t(\varphi),\Im_t\right)$ is a martingale.

Let condition (5.3) be fulfilled. Consider another probability measure $\widetilde{P}(d\omega) = \zeta_0^T(\varphi)P(d\omega)$ and denote by $\widetilde{E}\eta$ the expectation of the variable η with respect to the measure $\widetilde{P}(d\omega)$.

Lemma 5.3 *If condition* (5.3) *holds true, then for any $\Im_t$-measurable random variable η with $\widetilde{E}|\eta| < \infty$, we have*

$$\widetilde{E}\left[\eta/\Im_\tau\right] = \widetilde{E}\left[\eta\zeta_\tau^t(\varphi)/\Im_\tau\right], \quad P\text{-a.s.} \quad 0 \le \tau < t \le T.$$

The theorem below is the infinite-dimensional version of the Girsanov theorem.

Theorem 5.1 *Let* $\varphi = (\varphi(t), \Im_t) \in L_2(H, \Im_t)$ *and assume that condition* (5.3) *holds true. Then* $\xi = (\xi(t), \Im_t)$, *where*

$$\xi(t) = W(t) - \int_0^t \varphi(\tau)d\tau, \quad t \in [0, T],$$

is the Wiener process with respect to $\widetilde{P}$.

Proof Approximate $(\varphi(t), \Im_t)$ with a sequence of piecewise constant processes $(\varphi_n(t), \Im_t)$, such that $E\zeta_0{}^T(\varphi_n) = 1$. Denote

$$\xi_n(t) = W(t) - \int_0^t \varphi_n(\tau)d\tau, \quad t \in [0, T], \ \widetilde{P}_n(d\omega) = \zeta_0^T(\varphi_n)P(d\omega).$$

Taking into account Lemma 5.3, we obtain

$$\widetilde{E}_n[\exp\{(h, \xi_n(t) - \xi_n(\tau))\}/\Im_\tau] = \exp\left\{\frac{|h|^2}{2}(t-\tau)\right\}, \quad 0 \le \tau < t \le T,$$

where $h \in H_+$, and $\widetilde{E}_n$ denotes the expectation with respect to the measure $\widetilde{P}_n(d\omega)$. In other words, $(\xi_n(t), \Im_t)$ is the Wiener process with respect to the measure $\widetilde{P}_n(d\omega)$. Passing to the limit as $n \to \infty$, we obtain the statement of the theorem. □

The theorem below can be proved in the same fashion.

Theorem 5.2 *Let the process* $\varphi = (\varphi(t), \Im_t) \in L_2(H, \Im_t)$ *be such that*

$$\varsigma(t) = 1 + \int_0^t (\varphi(\tau), dW(\tau)), \quad t \in [0, T],$$

is a nonnegative continuous $(\Im_t, P)$*-martingale. Then* $(\xi(t), \Im_t)$, *where*

$$\xi(t) = W(t) - \int_0^t \frac{\varphi(\tau)}{\varsigma(\tau)}d\tau, \quad t \in [0, T],$$

is the Wiener process with respect to $\widetilde{P}(d\omega) = \varsigma(T)P(d\omega)$.

Theorem 5.1 naturally implies

Theorem 5.3 *Let* $(\xi(t), \Im_t)$ *be an Ito process and take the measurable process* $(\alpha(t), \Im_t)$ *such that*

$$P\left\{\int_0^T |\alpha(t)|^2 dt < \infty\right\} = 1 \tag{5.4}$$

and

$$E\exp\left\{\int_0^T (\alpha(u), dW(u)) - \frac{1}{2}\int_0^T |\phi(u)|^2 du\right\} = 1. \tag{5.5}$$

Then $\mu_\xi \sim \mu_W$, *and*

$$\frac{d\mu_W}{d\mu_\xi}(\xi) = E\left[\exp\left\{\int_0^T (\alpha(u), dW(u)) - \frac{1}{2}\int_0^T |\alpha(u)|^2 du\right\} \Big/ \Im_T^\xi\right]$$

P-almost surely, where $\Im_t^\xi = \sigma\{\xi(u), u \le t\}$, $t \in [0,T]$.

Proof It is easy to see that (5.5) holds true provided that for some $\delta > 0$ one has

$$E\exp\left\{\left(\frac{1}{2} + \delta\right)\int_0^T |\alpha(t)|^2 dt\right\} < \infty.$$

□

Remark 5.1 Condition (5.4) guarantees that $\mu_\xi \ll \mu_W$.

Thus, in the case of Ito processes we have only the sufficient conditions for $\mu_\xi \sim \mu_W$, and the derivative $\frac{d\mu_W}{d\mu_\xi}(\xi)$ can be calculated as the conditional expectation. For diffusion-type processes one can give the necessary and sufficient conditions for the equivalence of measures, and, moreover, the expressions for the densities are of much simpler form.

Lemma 5.4 *Let* $W = (W(t), \Im_t, P)$ *be a standard Wiener process on* $(\Omega, \Im, P)$ *with values in* H_-. *Then for any real random variable* ξ *which is measurable with respect to* $\Im_t^W = \sigma\{W(u), u \le t\}$, $t \in [0,T]$, *and* $E|\xi| < \infty$, *there exists a process* $\varphi = (\varphi(t), \Im_t^W) \in L_2(H, \Im_t^W)$ *such that*

$$\xi = \int_0^T (\varphi(u), dw(u)).$$

Proof Denote by $W_1(t)$, $W_2(t)$, ..., $W_n(t)$, ... the coordinates of the process $W(t)$ in the basis (e_i), and $\Im_T^{W_n} := \sigma\{W_n(u), u \in [0,T], i = 1,2,\ldots,n\}$, $\xi_n := E[\xi/\Im_T^{W_n}]$. Then ξ_n is the real square-integrable functional of the n-dimensional Wiener process $W^{(n)} = (W_1(t), W_2(t), \ldots, W_n(t))$, $t \in [0,T]$, such that $E\xi_n = 0$. Therefore, it admits the representation

$$\xi_n = \sum_{i=1}^{n} \int_0^T \varphi_{(i,n)}(t) dW_i(t),$$

where the processes $\left(\varphi_{(i,n)}, \Im_t^{W_n}\right)$, $t \in [0,T]$, $i = 1, 2, \ldots, n$, $n = 1, 2, \ldots$, satisfy

$$\sum_{i=1}^{n} \int_0^T E\left(\varphi_{(i,n)}(t)\right)^2 dt < \infty.$$

Put $\varphi_n(t) = \sum_{i=1}^{n} \varphi_{(i,n)}(t) e_i$. Then $(\phi_n(t), \Im_t^{W_n}) \in L_2(H, \Im_t^W)$, and

$$\xi_n = \int_0^T (\varphi_n(t), dW(t)).$$

Taking into account Property 3 of the stochastic integral we have

$$E\xi_n^2 = EI_T^2(\varphi_n) = \int_0^T E|\varphi_n(t)|^2 dt.$$

Since $\lim_{n\to\infty} \xi_n = \xi$, one can easily see that in the class $L_2(H, \Im_t^W)$ there exists a process $\phi = (\phi(t), \Im_t^W)$ such that

$$\lim_{n\to\infty} \int_0^T E|\varphi(t) - \varphi_n(t)|^2 dt = 0.$$

Therefore,

$$\xi = \lim_{n\to\infty} \xi_n = \lim_{n\to\infty} \int_0^T (\varphi_n(t), dW(t)) = \int_0^T (\varphi(t), dW(t)).$$ □

Lemma 5.5 *Let $W = (W(t), \Im_t, P)$ be a standard Wiener process on $(\Omega, \Im, P)$ with values in H_-. Then for any real random $\Im_t^W$-measurable variable ξ with $E|\xi| < \infty$, there exists a process $\phi = (\phi(t), \Im_t^W)$ such that for any $t \geq \tau$*

$$E\left[\xi/\Im_t^W\right]-E\left[\xi/\Im_\tau^W\right]=\int\limits_\tau^t(\varphi(u),dW(u))\ P\text{-a.s.}$$

The proof is based on Lemma 5.4, and follows the idea of the proof of Theorem 3 from [8].

Theorem 5.4 *Let* $(\xi(t),\Im_t)$ *be a diffusion-type process of the form* (5.1) *with* $\xi(0)=0$. *Then for* $\mu_\xi\sim\mu_W$ *it is necessary and sufficient that*

$$P\left\{\int\limits_0^T|\alpha(t,\xi)|^2dt<\infty\right\}=1$$

and

$$P\left\{\int\limits_0^T|\alpha(t,W)|^2dt<\infty\right\}=1.$$

In this case the respective densities are given by

$$\frac{d\mu_\xi}{d\mu_W}(W)=\exp\left\{\int\limits_0^T(\alpha(u),dW(u))-\frac{1}{2}\int\limits_0^T|\alpha(u)|^2du\right\},\quad P\text{-a.s.}\tag{5.6}$$

$$\frac{d\mu_W}{d\mu_\xi}(\xi)=\exp\left\{\int\limits_0^T(\alpha(u),dW(u))+\frac{1}{2}\int\limits_0^T|\alpha(u)|^2du\right\},\quad P\text{-a.s.}\tag{5.7}$$

The proof follows directly from Lemma 5.5 *and is similar to the proof of Theorem* 5 *from* [54].

As we see, the expressions for the densities $\frac{d\mu_W}{d\mu_\xi}(\xi)$ *and* $\frac{d\mu_\xi}{d\mu_W}(W)$ *are much more transparent comparing to the Ito case. Therefore, the question which naturally arises is the following: when an Ito process can be represented as a diffusion process?*

Theorem 5.5 *Let* $(\xi(t),\Im_t)$ *be an Ito process of the form* (5.2) *with* $\xi(0)=0$, *the process* $(\alpha(t),\Im_t)$ *is measurable and* $\int\limits_0^T E|\alpha(t)|dt<\infty$. *Put* $\overline{\alpha}(t,\xi)=E\left[\alpha(t)/\Im_t^\xi\right]$.

Then the random process $\nu=\left(\nu(t),\Im_t^\xi,P\right)$, *where* $\nu(t)=\xi(t)-\int\limits_0^t\overline{\alpha}(u,\xi)du$,

$t \in [0,T]$, *is the Wiener process and, consequently,* $(\xi(t), \Im_t)$ *is the diffusion-type process*

$$\xi(t) = \int_0^t \overline{\alpha}(u, \xi)du + \nu(t).$$

Proof Applying the Ito formula to the function $\exp\{i(h, \nu(t))\}$, $h \in H_+$, and taking the conditional expectation with respect to $\Im_u^\xi$, we get

$$E\left[\exp\{i(h, \nu(t) - \nu(u))\}/\Im_u^\xi\right] = \exp\left\{-\frac{|h|^2}{2}(t-u)\right\}$$

P-a.s, which implies that $\left(\nu(t), \Im_t^\xi, P\right)$ is the Wiener process.

Let $\xi(t)$ be an Ito field of the form (5.2), and suppose that the conditions of Theorem 5.4 are satisfied. Then, if

$$P\left\{\int_0^T |\overline{\alpha}(t,\xi)|^2 dt < \infty\right\} = P\left\{\int_0^T |\overline{\alpha}(t,W)|^2 dt < \infty\right\} = 1,$$

we get $\mu_\xi \sim \mu_W$, and the densities $\frac{d\mu_W}{d\mu_\xi}(\xi)$ and $\frac{d\mu_\xi}{d\mu_W}(W)$ are given by formulas (5.6) and (5.7) with $\overline{\alpha}(t,x)$ instead of $\alpha(t)$. In such a way, to calculate the densities it is necessary to calculate $\overline{\alpha}(t,\xi)$, or, in other words, to solve the filtration problem for the Ito process.

□

5.2 Filtration of Ito Processes in a Hilbert Space

We keep the notation of the previous section. Suppose that on $(\Omega, \Im, P)$ we have two stochastic processes $\eta = (\eta(t), \Im_t)$ and $\xi = (\xi(t), \Im_t)$. The process η takes its values in a complete separable metric space G, and its observations are not available. The observed process ξ is an Ito process with respect to $W = (W(t), \Im_t, P)$, $\xi(0) = 0$, and

$$\xi(T) = \int_0^T \alpha(s)ds + W(t), \quad t \in [0,T].$$

We assume that the stochastic process $\alpha(t)$ is measurable, adapted to the σ-algebras $\Im_t^{\eta W} = \sigma\{\eta(u), W(u), u \le t, \ t \in [0,T]\}$, and satisfying

$$\int_0^T E|\alpha(t)|^2 dt < \infty. \tag{5.8}$$

Note that since $\Im_t^{\eta W} \subseteq \Im_t$, we assume without loss of generality that $\Im_t^{\eta W} \equiv \Im_t$. Let f be a measurable on G functions, taking values in H, and satisfies

$$E|f(\eta(t))|^2 < \infty, \quad t \in [0,T] \tag{5.9}$$

Our problem is to estimate the function $f(\eta(t))$ from the observations of the process ξ in the interval $[0,t]$, using the least squares criterion. It is known that one can take as such an estimate the expression $E[f(\eta(t))/\Im_t^{\xi}]$. Motivated by this argument, we derive the stochastic differential equation for $E[f(\eta(t))/\Im_t^{\xi}]$, which is called the filtration equation. In what follows we write $E^t[\cdot] = E[\cdot/\Im_t^{\xi}]$.

Let us formulate our main assumption on the function f. We say that f belongs to the space $D(A)$, if there exists a measurable process $(A(t,f),\ t \in [0,T])$ taking values in H, adapted to the σ-algebras $\Im_t^{\eta\xi} = \sigma\{\eta(u), \xi(u), u \leq t,\ t \in [0,T]\}$, such that

$$\int_0^T E|A(t,f)|^2 dt < \infty,$$

and the stochastic process $(M(t,f),\ t \in [0,T])$, where

$$M(t,f) = f(\eta(t)) - f(\eta(0)) + \int_0^t A(u,f)du$$

is the $(\Im_t, P)$-martingale with values in H. Clearly, if $f \in D(A)$ and condition (5.9) holds true, then the process

$$\overline{M}(t,f) = E^t[f(\eta(t))] - E^0[f(\eta(0))] + \int_0^t E^u[A(u,f)]du, \quad t \in [0,T],$$

is the square integrable (with respect to the norm in H) $(\Im_t^{\xi}, P)$-martingale. We show that the martingale $(\overline{M}(t,f),\ \Im_t^{\xi})$ can be represented as the stochastic integral with respect to the Wiener process $(\nu(t), \Im_t^{\xi}, P)$ (from Theorem 5.5). We start with some preliminary statements.

Lemma 5.6 *Let $(W(t), \Im_t, P)$ be a standard Wiener processes on the probability space $(\Omega, \Im, P)$ taking values in H_-, and $(Z(t),\ t \in [0,T])$ is a separable square integrable $(\Im_t^W, P)$-martingale with values in H and zero expectation. Then there* *exists a stochastic process* ($\Phi(t)$, $t \in [0,T]$) *in the class* $L_2\left(L_2(H,H),\Im_t^W\right)$, *such that*

$$Z(t) = \int_0^t \Phi(u)dW(u), \quad t \in [0,T].$$

Proof It follows from Lemma 5.4 that any real-valued separable square integrable $\left(\Im_t^W, P\right)$-martingale ($\varsigma(t)$, $t \in [0,T]$) with zero expectation can be written as

$$\varsigma(t) = \int_0^t (\varphi(t), dW(t)), \quad t \in [0,T],$$

where ($\varphi(t)$, $t \in [0,T]$) is some process from the class $L_2\left(H, \Im_t^W\right)$. Applying this statement to the processes ($\varsigma_i(t)$, $t \in [0,T]$), $\varsigma_i(t) = (Z(t), e_i)$, $i = 1, 2, \ldots$, which are obviously real-valued separable square integrable $\left(\Im_t^W, P\right)$-martingales with zero expectations, we obtain that

$$\varsigma^i(t) = \int_0^t (\varphi_i(t), dW(t)), \quad t \in [0,T], \ i = 1, 2, \ldots,$$

where ($\varphi_i(t)$, $t \in [0,T]$) are the respective processes from $L_2\left(H, \Im_t^W\right)$.

Let $\Phi(t)$ be an operator with matrix representation ($\varphi_{ij}(t)$), $\varphi_{ij}(t) = (\varphi_i(t), e_j)$, $i, j = 1, 2, \ldots$, with respect to the basis (e_i). The function $\Phi(t)$ belongs to the class $L_2\left(L_2(H,H), \Im_t^W\right)$, and

$$Z(t) = \int_0^t \Phi(u)dW(u), \quad t \in [0,T]. \qquad \square$$

Lemma 5.7 *Assume that* $\xi = (\xi(t), \Im_t)$ *is an Ito process of the form* (5.2) *with* $\xi(0) = 0$, *and* $(\alpha(t), \Im_t)$, $\alpha \in L_2\left(H, \Im_t^W\right)$. *Then for any separable square integrable* $\left(\Im_t^\xi, P\right)$-*martingale* $\left(\widetilde{Z}(t), \ t \in [0,T]\right)$ *with values in H and zero expectation there exists a process* $\left(\widetilde{\Phi}(t), \ t \in [0,T]\right)$ *from the class* $L_2\left(L_2(H,H), \Im_t^\xi\right)$, *such that*

$$\widetilde{Z}(t) = \int_0^t \widetilde{\Phi}(u)d\nu(u), \quad t \in [0,T],$$

where $\left(\nu(t), \Im_t^\xi, P\right)$ *is a standard Wiener process from Theorem* 5.5.

Proof For simplicity, suppose that

$$E\exp\left\{-\int_0^T (\overline{\alpha}(t,\xi), d\nu(t)) - \frac{1}{2}\int_0^T |\overline{\alpha}(t,\xi)|^2 dt\right\} = 1.$$

For the above equality to hold true it is enough to assume that the process α is bounded in norm by a nonrandom constant. We apply Theorem 5.1 with $W(t) = \nu(t)$, $\varphi(t) = -\overline{\alpha}(t)$, $\Im_t = \Im_t^\xi$; according to this theorem, ξ is a Wiener process with respect to the probability measure $\widetilde{P}(d\omega) = \zeta_0^T(-\overline{\alpha})P(d\omega)$.

Put $Z(t) = \widetilde{Z}(t)\left[\zeta_0^t(-\overline{\alpha})\right]^{-1}$. It is easy to see that $(Z(t),\ t \in [0,T])$ is a $\left(\Im_t^\xi, P\right)$-martingale with values in H, which implies by Lemma 5.6 that in the class $L_2\left(L_2(H,H), \Im_t^\xi\right)$ there exists a stochastic process $(\Phi(t),\ t \in [0,T])$ such that

$$Z(t) = \int_0^t \Phi(u)d\xi(u) = \int_0^t \Phi(u)d\nu(u) + \int_0^t \Phi(u)\overline{\alpha}(u,\xi)du.$$

Since

$$\zeta_0^t(-\overline{\alpha}) = 1 - \int_0^t \zeta_0^u(-\overline{\alpha})(\overline{\alpha}(u,\xi), d\nu(u)),$$

we obtain, taking the product of stochastic integrals, that

$$\widetilde{Z}(t) = \int_0^t \widetilde{\Phi}(u)d\nu(u) + \int_0^t \widetilde{\varphi}(u)d\nu(u).$$

The process $\lambda(t) = \int_0^t \widetilde{\varphi}(u)d\nu(u)$ is (with probability 1) of bounded variation, and thus can be decomposed into a difference of two $\left(\Im_t^\xi, P\right)$-martingales. Thus, for any $t \in [0,T]$ we have $\lambda(t) = 0$ P-a.s., from where we derive the desired representation

$$\widetilde{Z}(t) = \int_0^t \widetilde{\Phi}(u)d\nu(u).$$

Therefore, by Lemma 5.7, in the class $L_2\left(L_2(H,H), \Im_t^\xi\right)$ there exists a stochastic process $\left(\widetilde{\Phi}(t),\ t \in [0,T]\right)$ such that

$$\overline{M}(t,f) = \int_0^t \widetilde{\Phi}(u)d\nu(u). \qquad \square$$

Theorem 5.6 *Assume that conditions* (5.8) *and* (5.9) *hold true,* $f \in D(A)$, *and*

$$\int_0^T E|f(\eta(t))|^2|\alpha(t)|^2 dt < \infty.$$

Then $E^t[f(\eta(t))]$ *satisfies the stochastic equation*

$$E^t[f(\eta(t))] - E^0[f(\eta(0))]$$
$$= \int_0^t E^u[A(u,f)]du + \int_0^t E^u[f(\eta(u) \otimes (\alpha(u) - \overline{\alpha}(u,\xi)) + D_u(f)]d\nu(u).$$

Here we denote by $f(\eta(u) \otimes (\alpha(u) - \overline{\alpha}(u,\xi))$ the tensor product of $f(\eta(u))$ and $\alpha(u) - \overline{\alpha}(u)$ which belong to the space H (by the tensor product $h_1 \otimes h_2$, $h_1, h_2 \in H$, we mean an operator that acts according to the rule $(h_1 \otimes h_2)h = h_1(h,h_2)$, $h \in H$), and the function $D_t(f)$ is for each $t \in [0,T]$ the Hilbert–Schmidt operator, whose matrix $(\alpha_{ij}(t))$ in the basis (e_i) is given by

$$\alpha_{ij}(t) = \frac{d}{dt}\langle M_i(f), W_j\rangle_t, \quad M_i(f) = (M_i(t,f), t \in [0,T]), \quad M_i(t,f) = (M(t,f), e_i).$$

Here $\langle M_i(f), W_j\rangle_t$ is the mutual structure function (see [28]) of the real-valued square integrable martingales $M_i(f) = (M_i(t,f),\ t \in [0,T])$ and $(W_i(t),\ t \in [0,T])$, $i, j = 1, 2, \ldots$

5.3 Controlled Stochastic Differential Equations in a Hilbert Space

Consider a stochastic differential equation of the type

$$d\xi(t) = f(t, \xi, u(t, \xi))dt + W(dt), \xi(0) = 0, t \in [0, T], \tag{5.10}$$

where $(W(t),\ t \in [0,T])$ is a Wiener process on some probability space $(\Omega, \mathfrak{I}, \mathrm{P})$ taking values in $H\_$, the control $u = (u_t,\ t \in [0,T])$ and the drift are non-anticipating functionals depending on the controlled process ξ and with values, respectively, in the space of controls U and in H. We need to find some control u^* in the space of admissible controls, such that u^* minimizes the criterion

$$E\int_0^T c(t,\xi)dt,$$

where the cost function c is some non-anticipating numeric functional.

First of all we clarify in which sense we understand the solution to (5.10). In order to have the existence of the strong solution to the stochastic differential equation (5.10), one should require the drift f, and hence the control u, to be sufficiently smooth. At the same time, the optimal functions are in general only measurable. In what follows, we give conditions which provide the existence and uniqueness of the so-called weak solution of (5.10) under any admissible control.

Let us introduce the necessary notation and the assumptions on the functions f and c. Let U be a compact set in some separable space, $\mathfrak{R}_U$ is the σ-algebra of Borel sets in U, U is the set of all $\mathfrak{R}_U \otimes \mathfrak{R}_T$-measurable non-anticipative (i.e., $\mathfrak{R}_t$-measurable for all t) functional $u_t(x)$, $t \in [0,T]$, $x \in C_T$, taking values in U ($\mathfrak{R}$ is the σ-algebra of Borel sets in $[0,T]$). The functionals belonging to U are called the admissible controls. The drift function appearing in (5.10) is supposed to satisfy the following conditions:

1. $f(t,x,u)$ is $\mathfrak{R} \times \mathfrak{R}_T \times \mathfrak{R}_U$-measurable, with values in H
2. For any $t \in [0,T]$ the function $f(t,\cdot,\cdot)$ is $\mathfrak{R}_t \times \mathfrak{R}_U$-measurable
3. For any $t \in [0,T]$ and $x \in C_T$ the function $f(t,x,\cdot)$ is continuous on U
4. For any $t \in [0,T]$ and $x \in C_T$ the set $f(t,x,U) = \{f(t,x,u),\ u \in U\}$ is convex and closed in H
5. There exists a constant $K > 0$ such that

$$\left|f(t,x,u)\right|^2 \le K\left(1 + \sup_{0\le s\le T} |x(s)|_-^2\right), t\in[0,T], x\in C_T, u\in U.$$

The cost function $c(t,x)$, $t \in [0,T]$, $x \in C_T$, is supposed to be $\mathfrak{R} \times \mathfrak{R}_t$-measurable, non-anticipative, and such that $0 \le c(t,x) \le L < \infty$.

Let $\alpha(t,x)$, $t \in [0,T]$, $x \in C_T$, be an $\mathfrak{R} \times \mathfrak{R}_t$-measurable non-anticipative functional with values in H.

Definition 5.3 We say that the stochastic differential equation

$$d\xi(t) = \alpha(t,\xi)dt + W(dt), t\in[0,T], \xi(0) = 0. \tag{5.11}$$

has a weak solution if there exists a probability space $(\Omega, \mathfrak{I}, \mathrm{P})$, a filtration $(\mathfrak{I}_t, t\in[0,T])$, $\mathfrak{I}_t \subset \mathfrak{I}$, a continuous stochastic process $\xi = (\xi(t), \mathfrak{I}_t)$, and a Wiener process $\xi = (\xi(t), \mathfrak{I}_t)$ with values in H_- such that

$$P\left\{\int_0^T |\alpha(t,\xi)|dt < \infty\right\} = 1,$$

and

$$\xi(t) = \int_0^t \alpha(s,\xi)ds + W(t).$$

P-almost surely for any $t \in [0,T]$.

Therefore, a weak solution is in fact the collection of objects $A = (\Omega, \Im, \Im_t, \mathrm{P}, W(t)\xi(t))$. For simplicity, the process $\xi = (\xi(t),\ t \in [0,T])$ will also be called the weak solution.

Denote by μ_ξ the measure on the space $(C_T, \Re_T)$ generated by the weak solution ξ. We say that the weak solution to (5.11) is unique if the measures corresponding to any two weak solutions ξ and $\widetilde{\xi}$ coincide. The following theorem gives the necessary and sufficient conditions for the existence and uniqueness of a weak solution to (5.11).

Theorem 5.7 *Assume that* $\alpha(t,x)$, $t \in [0,T]$, $x \in C_T$, *is a measurable non-anticipative functional with values in* H, *satisfying*

$$\int_0^T |\alpha(t,x)|^2 dt < \infty, x \in C_T. \tag{5.12}$$

Then for the existence of a weak solution to (5.11) *it is necessary and sufficient that there exists a Wiener process* $W' = \left(W'(t), \Im'_t, \mathrm{P}'\right)$ *on a probability space* $\left(\Omega', \Im', \mathrm{P}'\right)$ *with values in* $H\_$, *such that*

$$E' \exp\left\{ \int_0^T \left(\alpha\left(t, W'\right), W'(dt)\right) - \frac{1}{2}\int_0^T \left|\alpha\left(t, W'\right)\right|^2 dt \right\} = 1, \tag{5.13}$$

where E' *denotes the expectation with respect to the probability* P'.

Proof Necessity. Assume that a weak solution to (5.11) exists, that is, there exists a collection of objects $A = (\Omega, \Im, \Im_t, \mathrm{P}, W(t), \xi(t))$ such that for any $t \in [0,T]$

$$\xi(t) = \int_0^t \alpha(s,\xi)ds + W(t)$$

P-almost surely. Then by (5.12) we have

$$P\left\{ \int_0^T |\alpha(t,\xi)|^2 dt < \infty \right\} = 1,$$

and

$$P\left\{\int_0^T |\alpha(t,W)|^2 dt < \infty\right\} = 1.$$

Therefore, by Theorem 5.4 we have $\mu_\xi \sim \mu_W$, where μ_W is a Wiener on $(C_T, \Re_T)$, and

$$\frac{d\mu_\xi}{d\mu_W}(W) = \exp\left\{\int_0^T (\alpha(t,W), W(dt)) - \frac{1}{2}\int_0^T |\alpha(t,W)|^2 dt\right\}$$

P-almost surely. Then the conditions of the theorem are fulfilled for the Wiener process $W = (W(t), \Im_t, P)$ appearing in the weak solution ξ.

Sufficiency. Suppose that the assumptions of the theorem hold true. Then by Theorem 5.1 the process

$$\widetilde{W}(t) = W'(t) - \int_0^t \alpha\left(s, W'\right) ds, \quad t \in [0,T]$$

is the Wiener process with respect to the family of σ-algebras $\left(\Im'_t, t \in [0,T]\right)$ and the probability measure $\widetilde{P}(d\omega)$, where

$$\widetilde{P}(d\omega) = \exp\left\{\int_0^T \left(\alpha\left(t, W'\right), W'(dt)\right) - \frac{1}{2}\int_0^T \left|\alpha\left(t, W'\right)\right|^2 dt\right\} P'(d\omega).$$

This means that the collection $A = (\Omega, \Im, \Im_t, \mathrm{P}, W(t), \xi(t))$ defines the weak solution to (5.11). The uniqueness follows from the fact that the measure μ_ξ corresponding to any weak solution of (5.11) is equivalent to μ_W, and $\frac{d\mu_\xi}{d\mu_W}$ is defined in terms of the functional α only.

Let us come back to the stochastic differential equation (5.10). Put

$$g_u(t,x) - f(t,x,u(t,x)), t \in [0,T], x \in C_T, u \in U.$$

Property (5) of the drift function immediately implies that assumption (5.12) holds true. On the other hand, it follows from Lemma 5.7 from [31] that (5.13) holds for the functional g_u and an arbitrary Wiener process. Thus, for any admissible control $u = (u(t,x),\ t \in [0,T],\ x \in C_T) \in U$ there exists a unique weak solution to stochastic differential equation (5.11).

This solution can be constructed as follows. Fix some probability space $(\Omega, \Im, P_0)$ with a Wiener process $W = (W_t, \Im_t, P_0)$. Assume that $\Im_t = \Im_t^W = \sigma\{W_s, 0 \le s \le t\}$, $t \in [0,T]$, $\Im_T = \Im$ and complete the σ-algebra $\Im_0$ with respect to the measure P_0. Under these conditions, the flow of σ-algebras $(\Im_t t \in [0,T])$ is continuous. Let

$$D = \exp\{\varsigma_0^T(g_u), u \in U\}, \qquad P_u(d\omega) = \exp\{\varsigma_0^T(g_u)\}P(d\omega),$$

$$\varsigma_0^t(g_u) = \int_0^T (g_u(t,W), W(dt)) - \frac{1}{2}\int_0^T |g_u(t,W)|^2 dt, \quad t \in [0,T],$$

The process $\left(\widetilde{W}(t), \Im_t, \mathrm{P}_u\right)$, where

$$\widetilde{W}(t) = W(t) - \int_0^t g_u(s,W)ds, \quad t \in [0,T],$$

is the Wiener process, hence the collection of objects $A = (\Omega, \Im, \Im_t, P, W(t), \xi(t))$ gives the weak solution of (5.11). In such a way, the process $(W(t), t \in [0,T])$ considered with respect to the measure P_u is the weak solution to (5.11), which corresponds to any admissible control $u \in U$.

Then the problem described at the beginning of this section can be set up in the following way: find $u* \in U$ minimizing the functional

$$\Lambda(u) = E_u \int_0^T c(t,W)dt = E_0 \exp\{\varsigma_0^T(g_u)\} \int_0^T c(t,W)dt, \tag{5.14}$$

where E_u is the expectation with respect to P_u, $u \in U$.

Now let us show that the set D of densities is weakly compact in the space L_1 $(\Omega, \Im, P_0)$. We only need to prove that D is uniformly integrable and weakly closed in $L_1(\Omega, \Im, P_0)$.

Lemma 5.8 *There exists a constant* $\gamma * > 0$ *such that*

$$\sup_{u \in U} E_0 \exp\{\gamma * \varsigma_0^T(g_u)\} < \infty.$$

Proof We have

$$\exp\{\gamma\varsigma_0^T(g_u)\} = \exp\left\{\varsigma_0^T(\gamma g_u) + \frac{\gamma^2 - \gamma^T}{2}\int_0^T |f(t,W,u(t,W))|^2 dt\right\}$$

$$\le \exp\left\{\varsigma_0^T(\gamma g_u) + \frac{\gamma^2 - \gamma}{2}K\int_0^T \left(1 + \sup_{0 \le s \le t}|W(s)|^2\right)dt\right\}.$$

Put

$$\eta(t) = W(t) - \gamma \int_0^t g_u(s, W)ds.$$

Then

$$\left|W(t)\right|_-^2 \le 2\left|\eta(t)\right|_-^2 + 2\gamma^2 K(t) + \int_0^t \sup_{0\le v\le s} |W(v)|_-^2)dt,$$

and by the Gronwall–Bellman inequality

$$\sup_{0\le t\le T} \left|W(t)\right|_-^2 \le 2\left(\gamma^2 KT + \sup_{0\le t\le T} |\eta(t)|^2\right)\exp\{2\gamma^2 KT\},$$

whence

$$E_0 \exp\{\gamma\varsigma_0^T(g_u)\} \le h(\gamma)E_0 \exp\left\{\varsigma_0^T(\gamma g_u) + KT + e^{2\gamma^2 KT}(\gamma^2 - \gamma) \sup_{0\le t\le T} |\eta(t)_-^2\right\}$$

On the other hand, $E_0 \exp\{\gamma\varsigma_0{}^T(g_u)\} = 1$, and by Theorem 5.1 $W = (\eta(t), \Im_t, P')$ is the Wiener process, and $P'(d\omega) = \exp\{\varsigma_0{}^T(\gamma g_u)\}P_0(d\omega)$. Therefore,

$$E_0 \exp\{\gamma\varsigma_0^T(g_u)\} \le h(\gamma)E_0 \exp\left\{KTe^{2\gamma^2 KT}(\gamma^2 - \gamma) \sup_{0\le t\le T} |W(t)_-^2\right\}.$$

The function $h(\gamma)$ is bounded in the neighborhood of $\gamma = 0$. Take $\gamma > 1$ to be close enough to 1; then the expectation on the right-hand side of the last inequality is finite, and $E_0 \exp\{\gamma\varsigma_0{}^T(g_u)\} \le c_0 < \infty$, where the constant c_0 depends on γ, K, T only.

Lemma 5.9 *The set D is weakly closed in* $L_1(\Omega, \Im, P_0)$.

Proof Assume that the sequence of elements $\exp\{\gamma\varsigma_0{}^T(g_u)\}$ converges weakly to a random variable $f \in L_1(\Omega, \Im, P_0)$. It is clear that $E_0 f = 1$ and $f > 0$ with probability 1. We apply Lemma 5.4 to f as to an integrable functional of the Wiener process $W = (W(t), \Im, P_0)$. By this lemma, there exists a process $\varphi = (\varphi_t, \Im_t)$ with values in H, such that

$$P_0\left\{\int_0^T |\varphi(t)|^2 dt < \infty\right\} = 1, \tag{5.15}$$

and for any $0 \le s < t \le T$

$$E_0(f/\Im_t) - E_0(f/\Im_s) = \int_s^t (\varphi(u), W(du)) \tag{5.16}$$

P-almost surely.

Put $f_t = E_0(f/\Im_t)$. Since the family $(\Im_t, t \in [0,T])$ is continuous, one can deduce from the general theory of martingales that $P_0\{\inf_{0 \le t \le T} f(t) > 0\} = 1$. For the function

$$\psi(t) := \frac{\phi(t)}{f(t)} = \frac{\phi(t)}{1 + \int_0^t (\phi(s), W(ds))} \tag{5.17}$$

we have by (5.15) and the condition $P_0\{\inf_{0 \le t \le T} f(t) > 0\} = 1$ that

$$P_0\left\{\int_0^T |\psi(t)|^2 dt < \infty\right\} = 1.$$

By (5.16)–(5.17) we get

$$f(t) = 1 + \int_0^t f(s)(\psi(s), W(ds))$$

and

$$f(t) = \exp\{\varsigma_0^t(\psi)\}, f = f(T) = \exp\{\varsigma_0^T(\psi)\}.$$

For any $N > 0$ denote

$$\tau_N = \min(T, \inf t : |W(t)|\_N\},\ \chi_N = \chi_N(t, W) = \chi_{(\tau_N > t)},\ g_n^N = g_n \chi_N,\ \psi^N = \psi \chi_N.$$

The continuity of the family of σ-algebras $(\Im_t, t \in [0,T])$, of martingales $(\exp\{\varsigma_0{}^t(g_n)\})$, and of $(f(t))$ provides $f(\tau_N) = E_0(f/\Im_{\tau_N})$ and $\exp\{\varsigma_0^{\tau_N}(g_n)\} = E_0\left[\exp\{\varsigma_0^T(g_n)\}/\Im_{\tau_N}\right]$.

For any bounded random variable λ we have

$$E_0 \lambda E_0\left[\exp\{\varsigma_0^T(g_n)\}/\Im_{\tau_N}\right] = E_0 \exp\{\varsigma_0^T(g_n)\} E_0(\lambda/\Im_{\tau_N}),$$

$$E_0 \lambda E_0(f/\Im_{\tau_N}) = E_0 f E_0(\lambda/\Im_{\tau_N}).$$

Thus, for any $N > 0$

$$\exp\{\varsigma_0^T(g_n^N)\} = \exp\{\varsigma_0^{\tau_N}(g_n)\} \to \exp\{\varsigma_0^{\tau_N}(\psi)\} = \exp\{\varsigma_0^T(\psi^N)\},$$

weakly in $L_1(\Omega, \Im, P_0)$.

By condition (5) of the drift function f, the norms of the functions $g_n^N = g_n\ (t,W)\chi_N(t,W)$ are bounded by the same constant. Then the set of functions $\{\exp\{\varsigma_0{}^T(g_n{}^N)\}, n \geq 1\}$ is bounded in $L_2(H, \mathfrak{I}_t, P_0)$, and, consequently, is relatively compact in the weak topology of $L_2(H, \mathfrak{I}_t, P_0)$. This implies that $\exp\{\varsigma_0^T(\psi^N)\} \in L_2(\Omega, \mathfrak{I}, P_0)$ and $\exp\{\varsigma_0{}^T(g_n{}^N)\} \to \exp\{\varsigma_0{}^T(\psi^N)\}$ weakly in $L_2(H, \mathfrak{I}_t, P_0)$.

By the well-known result on the weak convergence, there exists a sequence of convex combinations $\left\{\sum_{k=1}^{k_n} \lambda_k^{(n)} \exp\{\varsigma_0^T(g_n^N)\}, n \geq 1\right\}$ strongly convergent to exp $\{\varsigma_0{}^T(\psi^N)\}$ in $L_2(H, \mathfrak{I}_t, P_0)$. One can easily see applying the Ito formula that

$$\sum_{k=1}^{k_n} \lambda_k^{(n)} \exp\{\varsigma_0^T(g_n^N)\} = \exp\{\varsigma_0^T(\eta_n),$$

where

$$\eta_n = \eta_n(t, W) = \sum_{k=1}^{k_n} \frac{\lambda_k^{(n)} \exp\{\varsigma_0^t(g_k^N)\}}{\sum_{k=1}^{k_N} \lambda_k^{(n)} \exp\{\varsigma_0^T(g_k^N)\}} g_k^N(t, W).$$

Since $g_n{}^N(t,W) = g_k(t,W)$, the inequality $\|W\|_- < N$ implies that $\eta_n(t,W) \in f(t, W(\omega), U)$ for $\|W\|_- < N$ (the set $f(t,x,U)$ is convex for any $t \in [0,T], x \in C_T$).

Therefore, we obtain

$$\lim_{n\to\infty} E_0 \left|\exp\{\varsigma_0^T(\eta_n)\} - \exp\{\varsigma_0^T(\psi^N)\}\right|^2 = 0.$$

whence from the Jensen inequality we get

$$\lim_{n\to\infty} E_0 \left|\exp\{\varsigma_0^t(\eta_n)\} - \exp\{\varsigma_0^t(\psi^N)\}\right|^2 = 0$$

for any $t \in [0,T]$. Applying the equalities

$$\exp\{\varsigma_0^T(\eta_n)\} = 1 + \int_0^T \exp\{\varsigma_0^t(\eta_n)\}(\eta_n(t)), W(dt)),$$

$$\exp\{\varsigma_0^T(\psi^N)\} = 1 + \int_0^T \exp\{\varsigma_0^t(\psi^N)\}(\psi^N(t)), W(dt)),$$

we obtain, using the properties of stochastic integrals, that

$$\lim_{n\to\infty} E_0 \int_0^T \left|\exp\{\varsigma_0^t(\eta_n)\}\eta_n(t) - \exp\{\varsigma_0^t(\psi^N)\}(\psi^N(t)\right|^2 dt = 0.$$

Thus, we have for some subsequence n_k

$$\lim_{n\to\infty} \exp\{\varsigma_0^t(\eta_{n_k})\} = \exp\{\varsigma_0^t(\psi^N)\},$$

$$\lim_{k\to\infty} \exp\{\varsigma_0^t(\eta_{n_k})\}\eta_{n_k} = \exp\{\varsigma_0^t(\psi^N)\}\psi^N(t),$$

$P_0 \otimes l$-almost everywhere, where l is the Lebesgue measure on $[0,T]$.

Since $\exp\{\varsigma_0{}^t(\psi^N)\}$ is positive, we obtain $\lim_{k\to\infty} \eta_{n_k} = \psi^N(t)$ $P_0 \otimes l$-almost everywhere.

On the other hand, $\eta_{n_k}(t, W) \in f(t, W(\omega), U)$ for $\|W\|_{\_} < N$. Therefore, since $f(t,x,U)$, $t \in [0,T]$, $x \in C_T$, is closed, we have $\psi(t, W(\omega)) \in f(t, W(\omega), U)$ almost everywhere with respect to measure $P_0 \otimes l$. By Lemma 5 from [4], there exists an admissible control $u \in U$ such that

$$\psi(t, W(\omega)) \in f(t, W(\omega), u(t, W(\omega))) = g_u(t, W(\omega)).$$ □

Therefore, the set D is weakly compact in the space $L_1(\Omega, \Im, P_0)$. The functional (5.14) is the continuous (in the topology of weak convergence in $L_1(\Omega, \Im, P_0)$) functional of the density $\exp\{\varsigma_0{}^t(g_u)\}$. Thus the following theorem holds true:

Theorem 5.8 *There exists $u^* \in U$, such that*

$$E_{u*}\int_0^T c(t, W)dt = \inf_{u \in U} E_u \int_0^T c(t, W)dt.$$

References

1. Achiezer, N.Y., Glasman, I.M.: Theory of linear operators in hilbert space. Nauka, Moscow (1966)
2. Bateman, H., Erdelyi, A.: Higher transcendental function. Graw Hill Book Company, New York–Toronto–London (1953)
3. Bazenov, L.G., Knopov, P.S.: On the forecast and filtration problems for random fields satisfying stochastic differential equations. News. Ac. Sci. USSR. Tech. Cybern. **6**, 53–157 (1974). In Russian
4. Benes, V.E.: Existence of optimal stochastic control laws. SIAM J. Control **9**(3), C446–C472 (1971)
5. Billingsley, P.: Convergence of probability measures. Wiley, New York (1968)
6. Brossard, J., Chevalier, L.: Calcul stochastique et inequalites de norm pour les martingales bi-browniennes. Application aux functions bi-harmoniques. Ann. Inst. Fourier. **30**(4), 97–120 (1980)
7. Cairoli, R., Walsh, J.B.: Stochastic integrals in the plane. Acta Math **134**, 111–183 (1975)
8. Clark J.M.S.: The representation of functionals of Brownian motion by stochastic integrals. Ann. Math. Statist. **41**(4), 1282–1295 (1970)
9. Daletzkiy Yu L. Infinite-dimensional elliptic operators and related parabolyc-type equations. Progr. of Math.Sci. **22**(4), 1–53 (1967) (In Russian)
10. Derieva, E.N.: Generalized girsanov theorem for two parameter martingales. Control and decision making methods under risks and uncertainty, pp. 43–50. V.M.Glushkov Institute of Cybernetics, Kiev (1993). In Russian
11. Derieva, E.N.: Stochastic equations of optimal nonlinear filtering of random fields. Cybernetics and system analysis **30**(5), 718–725 (1994)
12. Derieva, E.N.: Absolutely continuous change of measure in stochastic differential equations. Cybern Syst Anal **30**(6), 943–946 (1994)
13. Deriyeva O.M.: Control and identification problems for diffusion-type processes with distributed parameters. The international conference on Hanshan. Chernivtzi, Ruta, p. 39 (1994) (In Ukrainian)
14. Deriyeva O.M.: On some properties of measures corresponding to diffusion-type random fields on the plane. Krainian conference of Yang scientists. Mathematics, Kiev University, Kiev, 20 July 1994, pp. 253–260 (1994) (In Ukrainian)
15. Derieva O.M.: Some properties of measures corresponding to random fields of diffusion type on the plane. Theor. Probab. Math. Stat. **50**, 71–76 (1994)
16. Dorogovtzev A.Ya.: Remarks on the random processes generated by some differential equations. RAS. USSR. **3**(8), 1008–1010 (1962)
17. Dorogovtzev A.Ya., Knopov P.S.: Asymptotic properties of one parametrical estimate for a two-parameter function. Theor. Stoch. Proc. **5**, 27–35 (1977) (In Russian)

P.S. Knopov and O.N. Deriyeva, *Estimation and Control Problems for Stochastic Partial Differential Equations*, Springer Optimization and Its Applications 83, DOI 10.1007/978-1-4614-8286-4, © Springer Science+Business Media New York 2013

18. Ermoliev, Y.M., Gulenko, V.P., Tsarenko, T.I.: Finite difference approximation methods in optimal control theory. Naukova Dumka, Kiev (1978)
19. Etemadi N., Kallianpur G.: Nonanticipative transformations of the two-parameter wiener process and girsanov theorem. J. Multivariative Anal. **7**(1), 28–49 (1977)
20. Gihman, I.I.: Incremental two-parameter martingales. Proc. Seminar Stoch. Proc. Theor. Druskininkay. **1**, 35–69 (1975). In Russian
21. Gihman I.I.: Square integrable incremental two-parameter martingales. Probab. Theor. Math. Stat. (15), 21–30 (1976) (In Russian)
22. Gihman, I.I.: To theory of bi-martingales. Rep. Ac. Sci. Uk. SSR. **42**(6), 9–12 (1982) (In Russian)
23. Gihman, I.I.: Two-parameter martingales. Russian Math. Surveys. **37**(6), 1–30 (1982)
24. Gihman I.I., Piasetzkaya T.E.: Two types of stochastic integrals over martingale measures. Rep. Ac. Sci. Uk. SSR. **35**(11), 963–965 (1975) (In Russian)
25. Gihman I.I., Pyasetskaya T.E.: On a class of stochastic prtial differential equations containing two-parametric white noise. In book: Limit Theorems for Random Processes. Ac. Sci. Uk. SSR. Inst. Mat. 71–92 (1977) (In Russian)
26. Gihman, I.I., Skorochod, A.V.: Controlled stochastic processes. Springer, New York (1979)
27. Gihman, I.I., Skorochod, A.V.: Stochastic differential equations. Springer, Berlin (1972)
28. Gihman, I.I., Skorochod, A.V.: The theory of stochastic processes, vol. 1. Springer, New York (1979)
29. Gihman, I.I., Skorochod, A.V.: The theory of stochastic processes, vol. 3. Springer, New York (1979)
30. Gihman, I.I.: A generalization of inverse kolmogorov equation. The behavior of systems in random environments, pp. 9–17. Institute of Cybernetics, Ukrainian Academy of Sciences Uk SSR, Kiev (1977). In Russian
31. Gihman, I.I.: On a generalization of inverse kolmogorov equation. The behavior of systems in random environments, pp. 21–27. Institute of Cybernetics, Ukrainian Academy of Sciences Uk SSR, Kiev (1979). In Russian
32. Gihman I, I.I.: On a quasilinear partial stochastic differential equations. Theor. Stoch. Proc. **5**(21–27) (1977)
33. Girsanov, I.V.: On transforming a certain class of stochastic processes by absolutely continuous substitution of measures. Theor. Probab. Appl. **5**(3), 285–301 (1960)
34. Gubenko L.G., Knopov P.S., Statland E.S.: On a control problem for quasi-diffusion processes. Theor. Optim. Solut. **2**, 98–102 (1969) (In Russian)
35. Gushchin, A.A.: On the general theory of random fields on the plane. Russian Math. Surveys. **37**(6), 55–80 (1982)
36. Gyon, X., Prum, B.: Le theorem Girsanov pour an classe de processes a parameter multidimensional. G. R. Acad. Sc. Paris. **285**, 565–567 (1977)
37. Kaliath T.: A note on least squares estimation by the innovations method. SIAM J. Control. **10**(3), 477–486 (1972)
38. Kallianpur, G.: Stochastic filtering theory. Springer, Berlin (1980)
39. Knopov P.S.: To the forecast problem for random fields. Theor. Optim. Solut. **3**, 74–83 (1968) (In Russian)
40. Knopov P.S.: On one control problem for markov field satisfying a stochastic differential equation. Theor. Optim. Solut. **3**, 109–115 (1969) (In Russian)
41. Knopov P.S.: Filtering and prediction problems for random functions. Cybernetics. **8**(6), 953–978 (1972)
42. Knopov, P.S.: Optimal estimators of parameters of stochastic system. Naukova Dumka, Kiev (1981). In Russian
43. Knopov, P.S.: On an estimator for the drift parameter of random fields. Theor. Probab. Math. Stat. **45**, 29–33 (1992)
44. Knopov, P.S.: Some applied problems from random field theory. Cybern. Syst. Anal. **46**(1), 62–71 (2010)

45. Knopov, P.S., Derieva, O.M.: A generalized Girsanov theorem and its statistical applications. Theor. Probab. Math. Stat. **51**, 69–79 (1995)
46. Knopov, P.S., Derieva, E.N.: Control problem for diffusion-type random fields. Cybern. Syst. Anal. **31**(1), 52–64 (1995)
47. Knopov, P.S., Statland, E.S.: On absolutely continuous measures corresponding to some stochastic fields on the plane. Statistic and control problems for random processes, pp. 158–169. Institute of Cybernetics, Ukrainian Academy of Sciences Uk SSR, Kiev (1973). In Russian
48. Knopov P.S., Statland E.S.: On hypothesis distinction for some classes of stochastic system with distributed parameters. Cybernetics. **10**(2), 106–C107 (1974) (In Russian)
49. Knopov, P.S., Statland, E.S.: On one control problem for a solution of stochastic differential equation in hilbert space (Prepr. 75-29), p. 19. Institute of Cybernetics, Ukrainian Academy of Sciences Uk SSR, Kiev (1975). In Russian
50. Knopov, P.S., Statland, E.S.: On control of solution of stochastic differential equation in Hilbert space. Teor. Jmovirn. Mat. Stat. **65**, 53–59 (2001)
51. Krylov, N.V.: Controlled diffusion processes. Springer, Berlin–Heidelberg (1980)
52. Ledoux, M.: Inegalites de Burkholder pour martingales indexes par N×N. Lect. Notes. Math. **863**, 122–127 (1981)
53. Lions, J.-L.: Optimal control of systems governed by partial differential equations, p. 396. Springer, Berlin (1971)
54. Liptser, R.S., Shiriaev, A.N.: Statistics of random processes: General theory. Springer, New York (2001)
55. Metraux, C.: Qualques integralites pour martingales a parameter bidimensionnel. Lect. Notes. Math. **649**, 170–179 (1977)
56. Meyer, P.-A.: Probability and potentials. Blaisdell Publishing Co., New York (1966)
57. Mishura, Y.S.: A generalized Ito formula for two-parameter martingales II. Theor. Probab. Math. Stat. **32**, 77–94 (1986)
58. Mishura, Y.S.: Decomposition of two-parameter martingales into orthogonal components. Theor. Probab. Math. Stat. **23**, 127–136 (1981)
59. Mishura, Y.S.: Ito's formula for two-parameter stochastic integrals with respect to martingale measures. Ukr. Math. J. **36**(4), 370–374 (1984)
60. Mishura, Y.S.: Stochastic differential equations in the plane that contain strong semimartingales. Theor. Probab. Math. Stat. **45**, 77–85 (1992)
61. Mishura Yu.S. A Martingale Characterization of Diffusion Random Fields on the Plane. Theory of Probability & Its Applications. – 1990. - Vol. 35, No. 1. - P. 152-157.
62. Mishura, Y.S.: On the measure changing for two-parameter processes. Theor. Probab. Appl **30**(3), 612–613 (1985)
63. Novikov, A.A.: Optimal interpolation of partly observed two-parameter random fields—Probability theory, vol. 5, pp. 211–220. Banach Center Publications, PWN—Polish scientific Publishers, Warsaw (1979)
64. Piacetzkaya T.E.: Stochastic integration on the plane and one class of stochastic systems of hyperbolic differential equations. Thesis Ph.D.in Math, Donetzk (1977) (In Russian)
65. Ponomarenko, L.L.: Linear filtering of random fields controlled by stochastic equations. Cybern. Syst. Anal. **9**(2), 287–292 (1973)
66. Agarwal, R.P., Kim, Y.-H., Sen, S.K.: Multidimensional Gronwall-Bellman-type integral inequalities with applications. Memoir. Differ. Equat. Math. Phys. **47**, 19–122 (2009)
67. Rozovsky, B.L.: Stochastic evolution systems. Kluwer Academic, Dordrecht (1990)
68. Shurko, G.K.: On the convergence of one class of two-parameter stochastic integral equations. Theor. Stoch. Proc. **16**, 97–105 (1988). In Russian
69. Sokolovskiy, V.Z.: Kalman filters for distributed systems (Prepr. 74-60). Institute of Cybernetics, Ukrainian Academy of Sciences Uk SSR, Kiev (1974). In Russian
70. Statland, B.S., Statland, E.S.: Absolutely continuity and equivalence of measures corresponding to certain stochastic processes with values in Hilbert space, and filtering

problems. Theory of random processes: Problems of statistics and control, pp. 238–256. Institute of Mathematics, Kiev (1974). In Russian
71. Tzarenko, T.I.: The existence and uniqueness of solutions of stochastic Darbou equations. Theory of optimal solutions. Institute of Cybernetics, Kiev (1973). In Russian
72. Wong, E., Zakai, M.: Martingales and stochastic integrals for processes with a multidimensional parameters. Z. Wahrscheinlichkeitstheorie. Verw. Geb. **291**, 109–122 (1974)

Index

P.S. Knopov and O.N. Deriyeva, *Estimation and Control Problems for Stochastic Partial Differential Equations*, Springer Optimization and Its Applications 83,
DOI 10.1007/978-1-4614-8286-4, © Springer Science+Business Media New York 2013

U

W

Printed in the United States
By Bookmasters

Printed in the United States
By Bookmasters